SEP 1 5 2010

D0966315

THE
YOUTH
PILL

CURRENT

THE YOUTH PILL

SCIENTISTS AT THE BRINK OF AN
ANTI-AGING REVOLUTION

DAVID STIPP

CURRENT

CURRENT
Published by the Penguin Group
Penguin Group (USA) Inc., 375 Hudson Street,
New York, New York 10014, U.S.A.
Penguin Group (Canada), 90 Eglinton Avenue East, Suite 700,
Toronto, Ontario, Canada M4P 2Y3
(a division of Pearson Penguin Canada Inc.)
Penguin Books Ltd, 80 Strand, London WC2R 0RL, England
Penguin Ireland, 25 St. Stephen's Green, Dublin 2, Ireland
(a division of Penguin Books Ltd)
Penguin Books Australia Ltd, 250 Camberwell Road, Camberwell,
Victoria 3124, Australia
(a division of Pearson Australia Group Pty Ltd)
Penguin Books India Pvt Ltd, 11 Community Centre, Panchsheel Park,
New Delhi – 110 017, India
Penguin Group (NZ), 67 Apollo Drive, Rosedale, North Shore 0632,
New Zealand (a division of Pearson New Zealand Ltd)
Penguin Books (South Africa) (Pty) Ltd, 24 Sturdee Avenue,
Rosebank, Johannesburg 2196, South Africa

Penguin Books Ltd, Registered Offices:
80 Strand, London WC2R 0RL, England

First published in 2010 by Current,
a member of Penguin Group (USA) Inc.

10   9   8   7   6   5   4   3   2   1

LIBRARY OF CONGRESS CATALOGING IN PUBLICATION DATA
Stipp, David.
    The youth pill : scientists at the brink of an anti-aging revolution / David Stipp.
        p. cm.
    Includes bibliographical references and index.
    ISBN 978-1-61723-000-4
    1. Longevity—Popular works. I. Title.
    QP85.S75 2010
    612.6'8—dc22   2010007114

Printed in the United States of America
Set in Whitman
Designed by Pauline Neuwirth

To Alicia

The idea is to die young as late as possible.

—ASHLEY MONTAGU (1905–1999)

CONTENTS

THE
YOUTH
PILL

THE MOUSE STUDY
THAT ROARED

A FEW HOURS before one of the biggest medical stories of 2006 broke, David Sinclair, the man who initiated the research, was addressing about half a million people from the driver's seat of his Honda Accord—a *Financial Times* reporter was interviewing him via a speakerphone in the car for a story that would run the next day. It was 8:15 A.M. on a freakishly balmy November day in Boston, and the Harvard Medical School scientist was in his many-armed-deity mode, dodging and weaving through rush-hour traffic, sipping his second tall black coffee of the day, stealing glances at incoming e-mail on his handheld Treo, and, in a routine he'd repeated so many times it was almost effortless, supplying the reporter with hooks to snag even the most science-shy readers. As usual during his morning commute, it occurred to Sinclair that the Riverway, a winding parkway he took to work, was a bad place for mobile multitasking—too many close-packed commuters veering hell-for-leather around tree-lined curves. But the thought quickly disappeared over his mental horizon. Those of many arms don't wrap cars around trees. Besides, he had to stay focused.

Sinclair finished the interview while pulling into a parking lot at the New Research Building, a glass tower grafted into Harvard's venerable med-school campus like a titanium implant in an octogenarian's hip. After gulping the last of his coffee, he dropped the paper cup on his car floor, where it joined a dozen others, hurried up to his ninth-floor lab, and stepped onto the runaway tilt-a-whirl of a Warholian fifteen minutes of fame.

Since getting to work at 6:30, Susan DeStefano, the Sinclair lab's unflappable coordinator, had continually updated her boss's schedule for the day as reporter after reporter called to plead for interviews. European media would get callbacks first because deadlines were already looming across the Atlantic. As Sinclair began reeling off sound bites over the phone, DeStefano handed him a cup of his favorite Italian roast. It turned out to be his last coffee of the day, leaving him three cups short of his usual intake. He wound up doing back-to-back interviews and photo shoots for the next nine hours straight, surfacing late in the day to realize with surprise that he had been going nonstop without his customary caffeine jolts, much less anything to eat.

There was some comic relief during the maelstrom. At an afternoon photo shoot for the *New York Times*, Sinclair found himself trying to look appropriately Harvardish while two lab mice that the photographer had asked him to pose with had scurried up his arm and tried to hide inside his shirt collar. As he later rushed into a Boston TV studio to appear on *The Charlie Rose Show*, DeStefano, who was right behind him, noticed that Sinclair's spiky hair was sticking out like a horizontal cowlick. Frantically improvising, she spit on her hand and smoothed down the mischievous sprig. Somehow it stayed.

The study that caused the ruckus, published that day by the journal *Nature*, wasn't at first glance the kind of biomedical starburst that flings front-page headlines across the globe. Its pedestrian title, "Resveratrol Improves Health and Survival of Mice on High-Fat Diets," suggested a minor finding that might, on a slow news day, generate blurbs in a few newspapers. Its subjects were only rodents, so the implications for human health were still iffy. And what was resveratrol anyway, a new diet pill whose name was a mangled pun on "reverse-it-all"?

But something so remarkable had happened to the study's mice that even the media's gruffest gatekeepers were compelled to suspend disbelief about its newsworthiness. Daily doses of resveratrol, a compound found in red wine, had rendered the rodents largely immune to fallout from devastatingly rich diets they'd been given during the study, suggesting that the compound might be highly effective at fending off the array of life-shortening ills coming at us during the era of epidemic obesity. Even more arresting, the animals had shown amazingly youthful vigor as they had aged, as if their fundamental life force had been strengthened and their aging process slowed down.

I was one of the first outsiders to hear about the study—as a medical reporter at the *Wall Street Journal* and *Fortune* over the previous two decades, I'd gotten hooked on aging science and followed it more closely than any other topic I covered. Sinclair was one of the researchers on aging I often talked to, and months earlier he'd told me his group was beginning to see interesting effects in mice on resveratrol. Given all the similar claims that have led nowhere, I generally have only one question about the discovery of interventions that purportedly extend life span: Why should I buy this? The answers rarely whet my appetite to hear more. But the resveratrol study had more going for it than any revelation of anti-aging effects I'd seen.

First, the work was blessed by *Nature*, widely regarded as the world's top science journal, and its authors worked at prestigious institutions. While Sinclair had commercial ties, the study hadn't received corporate funding and was largely conducted by researchers at the National Institute on Aging. Further, resveratrol had previously been reported to extend life span in yeast, flies, worms, and fish—no other compound had shown anti-aging effects in such a diverse array of species. Perhaps most important, there was growing evidence that resveratrol mimics effects of calorie restriction, or CR, a curtailment of food intake that had been known for more than seventy years to extend life span in animals. That offered a plausible explanation for its apparent anti-aging effects and tied the study's results to firmly established prior research.

As the study progressed, the preliminary data it generated seemed ever more intriguing. Several months before the *Nature* report was published, I visited the federal lab in Baltimore where the resveratrol-dosed rodents were kept—I wanted to see for myself the remarkable effects Sinclair had told me about. After watching elderly mice on resveratrol perform like rodent Olympians in an endurance test, I came away convinced that the long, weird quest to extend life span—a five-thousand-year trek during which hopelessly hopeful seekers tried everything from transfusing blood from youths into their aged veins to injecting minced dog testicles—was finally getting somewhere. It also struck me that substances capable of inducing CR's effects wouldn't simply prolong a miserable old age—they would probably buy us quality time.

Inevitably, there were caveats, complexities, and questions. The mice had been given huge doses of resveratrol. A person attempting to get a comparable dose would need to knock back whole cases of wine a day. And there was no guarantee that resveratrol would have anti-aging effects in people—the provocative discovery might lead to nothing more than pep pills for bloated rodents.

Still, it was the mouse study that roared. Weeks before it was published, reporters with the *New York Times* and other news outlets got wind of its findings and called Sinclair to ask about it. A few days before it was widely reported, *Nature*'s editors, realizing they had a hot paper in their hands, alerted media outlets that they would unveil the study online almost two weeks before it appeared in the British science journal's print version. Soon after, Sinclair found himself dealing with the kind of early morning calls that newly anointed Nobel laureates wake up to. One eager reporter called him just after seven A.M., when, as usual, the Harvard scientist was overseeing his toddler daughters' breakfast after his wife left for work. Acutely aware of the media's short attention span, Sinclair assured the importunate writer that okay, sure, it was a fine time to talk. In a minor stroke of genius, he kept the kids somewhat quiet during the interview by plying them with hastily slapped-together Nutella sandwiches. To the girls, Sinclair's big new thing was redolent of the chocolaty spread. For their dad, it had the even sweeter smell of vindication.

Eleven years earlier, Sinclair had earned a Ph.D. in biology, and then, in a move that baffled his mentors and peers, plunged into gerontology, the study of aging, at a time when most biologists regarded the field as barren ground. The process of aging, ran the conventional wisdom, was too chaotic and intractable to offer much of interest to serious scientists. But to Sinclair and a small number of like-minded biologists, the time seemed ripe to rejuvenate gerontology, bringing the powerful new tools of molecular biology to bear on its central mysteries. Instead of simply describing aging's myriad curiosities, they would prize them apart molecule by molecule. Some dared to hope that even before their careers had wound down, their work might lead to drugs that would shrink death's dominion more than any medical advance ever had.

Today that goal is looming up a lot faster than anyone expected a decade ago. Even the famously conservative pharmaceutical industry, which generally shies away from radically new research frontiers, has gotten involved. Two years after Sinclair's mouse study, GlaxoSmith-Kline, the world's second biggest drug company, shelled out a stunning $720 million for Sirtris Pharmaceuticals, a biotech start-up the Harvard researcher cofounded in 2004 with prominent venture capitalist Christoph Westphal to apply his anti-aging discoveries.

Major advances in aging science have been unfolding so fast that even some gerontologists can scarcely believe that the once-distant goal is now clearly within reach. Rafael de Cabo, a researcher at the National Institute on Aging who collaborated with Sinclair on the resveratrol study in mice, told me in 2006, "I confessed to David before we started the study that I was a nonbeliever. I didn't think resveratrol would work in mammals" and induce longevity-enhancing effects. "I warned him that we would publish the findings regardless of what they were. He just said, 'That's fine. I have full confidence it will work.'"

In the normal course of things, Sinclair's breezy rejoinder would have been famous last words—the gods would have struck down the Harvard scientist for first-degree hubris, and the mice would have keeled over in a trice after being fed resveratrol. But the normal course of things in the search for ways to delay time's toll has ended. Sinclair is still very much with us, as is gerontologists' sense that their field is finally in a position to develop medicines that truly slow aging. This book is the story of how it all happened.

1

THE WILD OLD WICKED MYSTERY OF AGING

THE "STRANGE SECOND puberty" of William Butler Yeats, as he called it, began in April 1934 when he traveled to London at age sixty-eight for a fifteen-minute surgical procedure that induced one of history's most sensational placebo effects. Basically a vasectomy, the "Steinach operation" Yeats underwent was widely believed to induce profound rejuvenating effects, beginning with restoration of youthful sexual powers. It worked wonders for the high-strung, suggestible poet. Soon after the operation he enthused to his publisher that it "has almost made me a new man. . . . I no longer feel myself at the end of life putting things in order or putting them away."

Over the next five years, Yeats re-created himself as the "wild old wicked man" depicted in one of his late poems. Despite being chronically afflicted with various ailments, including impotence, the portly near-septuagenarian flung himself into a series of fervid affairs with much younger women. His muse, always given to consorting with Eros, caught fire one last time. Between 1936 and 1939, the last three years of his life, he wrote fifty poems, including some of his finest, compared to just seventy during the entire 1920s. Yeats's late, great surge left no doubt that he then was, and would very likely remain, the century's greatest English-language poet.

His medically induced rejuvenation also elicited ridicule. Dublin newspapers dubbed him "the gland old man." A younger Irish writer joked that recharging the aging poet's priapic batteries was "like putting a Cadillac engine into a Ford car." Even one of his friends couldn't resist penning a cruel parody of one of his poems—"I heard the old, old men say/'Everything alters'" became "I heard the old, old men say/'Everything's phallic.'"

Yeats's operation, named for and devised by pioneering Viennese endocrinologist Eugen Steinach, was thought to stimulate proliferation of gonadal cells whose secretions could reactivate the generalized vigor of puberty. The Irish poet wasn't the only celebrity eager to believe it worked. Sigmund Freud had had the operation in 1923 at age sixty-seven in hopes that it would prevent recurrence of his oral cancer, perk up his sex life, and boost his capacity for work. After undergoing a female version of the procedure, involving the irradiation of her ovaries, sixty-six-year-old American novelist Gertrude Atherton claimed she felt thirty years younger.

It was all wishful thinking. But there was more to it than the tomfoolery of faded peacocks. The renewal that Steinach and others promised in the wake of World War I and the terrible "Spanish flu" pandemic of 1918 resonated with deep yearnings—turning on inner fountains of youth seemed just what the doctor ordered for depleted humanity. Two years after Oswald Spengler published *The Decline of the West* in 1918, Steinach brought out his *Rejuvenation Through the Experimental Revitalization of the Aging Puberty Gland*. Steinach's admirers nominated him for a Nobel Prize no less than six times after his rejuvenation book came out, though he never got one. By the time Yeats had himself "Steinached," untold thousands of wildly hopeful old men across the globe had also had the operation.

The rage for gonadally-based rejuvenation inspired stranger things too. In 1889, Charles-Édouard Brown-Séquard, a prominent French neurologist, had reported experiencing rejuvenating effects after injecting himself with minced dog and guinea pig testicles. During the 1920s and 1930s, French-Russian surgeon Serge Voronoff and his imitators went further, implanting testicular grafts from apes and other animals in many thousands of aging men. Voronoff even set up a

monkey farm in Italy to meet soaring demand for gonads. One enthusiastic U.S. experimenter, L. L. Stanley, staff physician at California's
San Quentin State Prison, reported in 1922 that he'd placed testicular
implants from various kinds of animals into 656 research subjects,
including inmates and even women. The surgeries, he claimed, had
mitigated everything from paranoia to senility. Stanley also listed
eleven dead people among his subjects—always the painstaking
empiricist, he duly noted that none of the corpses had benefited. The
most notorious gland man was John Brinkley, a Kansas evangelist and
"physician" who made a fortune installing goat testicles in more than
sixteen thousand true believers who flocked to his "medico-gospel"
farm. An unknown number of his patients reportedly died from infections and other complications of the surgery, which he sometimes
performed while inebriated.

The great gland madness of the late nineteenth and early twentieth
centuries marked the first full-fledged world craze to spring from the
application of modern science to the anti-aging quest. The craziness
lasted for decades despite the fact that its leaders were often blasted
by skeptics and mocked by satirists. (Even the Marx Brothers got into
the act—an Irving Berlin song titled "The Monkey Doodle Doo" from
their 1929 film *The Cocoanuts* includes the lyrics "If you're too old for
dancing/Get yourself a monkey gland.") Unfortunately, its fallout continued long after its practitioners had been discredited.

The worst side effect was the dark cloud it stationed over serious
research on aging. Respected physician-scientists whose work helped
foster the rise of modern endocrinology, such as Brown-Séquard and
Steinach, had not only helped to instigate the most lurid case of mass
quackery in history, they had been laughably wrong about what drives
the aging process, showing how easily medical authorities can delude
themselves on the subject. The sex-gland affair was by no means the
only wacko, pseudoscientific fad of the past century that suggested
research on aging was mainly for fools and charlatans. But it was by
far the most memorable one.

The nature of aging has always been one of biology's deepest and
richest subjects, however, making it a magnet for brilliant as well as
misguided investigators. Even before the gland craze ended, pierc-

ing minds had begun to unravel gerontology's fundamental mystery: Why do we get old? A decade after Yeats's operation, a young British researcher came up with the answer. His was the first theory on aging that aged well.

. . .

Eccentricity springs eternal at Oxford University, but the sight of one of its young dons carrying rabbits across school grounds during the waning months of World War II struck some school officials as deplorably undignified. It wasn't the first time Peter Medawar had prompted whispers and frowns beneath the dreaming spires. Tall, slim, and rakishly handsome—his prominent, arched eyebrows gave him a permanent look of faint, diabolical amusement—the Brazilian-born biologist regarded Oxford's unreconstructed Victorians as irresistible targets for his formidable wit. He was the son of a Lebanese businessman who had once worked for a dental-supply company in Rio de Janeiro, and when asked by fellow Oxonians what his father did, his standard reply was, "He sells false teeth in South America." Medawar confessed years later that "it was a joy to see the agonies of embarrassment" this answer caused his upper-class interlocutors, who often heightened his amusement by struggling to find words "to relieve me of the burden of shame which they took for granted must weigh upon my every waking moment."

Such jaunty irreverence might have been ruinous for a lesser talent. Many of the claret-sipping sages Medawar rubbed elbows with, such as C. S. Lewis and J. R. R. Tolkien, were born during the reign of Queen Victoria. But his outsized intellect easily outshone the mischievous streak. As an Oxford undergraduate a decade earlier, he'd stood out as a bohemian dandy, sporting cheap jackets that were often overdue for a cleaning, cracking jokes about padding his worn-out shoes with newspaper, and treating much of his coursework as optional. Shortly before graduation he flippantly asked a classmate, his future wife, "Would you like me to get a First?"—a top-honors degree. When she said yes, he spent the next few weeks furiously reading up on his major, zoology. "Much of the information came as a great surprise to me," he drily commented in his autobiography. But he handily won the coveted degree.

As a junior researcher at Oxford a few years later, Medawar initiated immunology studies that earned him a Nobel Prize in 1960 at age forty-five. The rabbits he'd been spotted lugging across the quads were his experimental subjects. He was investigating why bits of skin from one animal quickly shriveled when grafted onto another one—it was desperately necessary to understand and overcome skin-graft rejection in order to help treat wartime burn victims. A prodigious worker, Medawar performed hundreds of skin grafts on his rabbits while working out how the immune system attacks "nonself" tissues, a prelude to successful organ transplants. He even cleaned his rabbits' cages. But somehow he found time to pursue other interests, including carrying out elaborate thought experiments on aging.

Medawar made light of his foray into gerontology, casually noting later that he "devoted a little of [his] time" to it during the 1940s. But he was proud of his insights, noting that they were well received by renowned biologists such as Britain's J. B. S. Haldane. (Haldane's approval was especially important because the famously eccentric genius was known for sitting in the front row of lectures by scientists he disliked, holding his head in his hands, and exclaiming "Oh God, Oh God!" in a penetrating voice.) Medawar later gave pride of place to the two seminal essays he wrote on aging by opening his first book with them. When named in 1951 to England's most venerable chair in zoology at the University of London, the thirty-seven-year-old "boy professor" presented his aging theory as an inaugural lecture at the school under the title "An Unsolved Problem of Biology"—implying, of course, that the solution was at hand.

Medawar didn't singlehandedly polish off aging science's "why" question. His ideas were foreshadowed in the works of other scientists—Haldane was one—and evolutionary biologists would later refine his theory. But discussions about the basic nature of aging are now shot through with his key idea. It holds that the blind watchmaker, to borrow Richard Dawkins's witty term for evolution's fitness-boosting process of natural selection, essentially loses interest in us after the age of reproduction and abandons us to the ravages of time.

Medawar was the first to spell out a lucid, detailed version of this concept, which stands as one of gerontology's cornerstones.

Before the mid-twentieth century, gerontologists typically spent their careers compiling minutiae on their favorite aspects of aging and, like the scientists who saw gonadal decay as the driver of senescence, proposing loopy but lucrative ways to stop or reverse them. As a result, the field's subject matter was little more than a bewildering catalog of physical changes vaguely correlated with the aging process. It held a certain morbid fascination. As early as 1932, for example, it was known that brain weight peaks around age twenty, then declines slowly until one's midfifties, after which the organ shrinks like a bowl of Jell-O left out in the hot sun. But the gerontology literature mainly inspired anxious self-examination and dubious nostrums. Then Medawar nonchalantly walked by with a rabbit under his arm and pointed the way out of the swamp.

His timing was perfect. Since the 1930s, a cadre of brilliant evolutionary thinkers—Haldane, R. A. Fisher, Sewall Wright, and others—had been assembling a grand, mathematically rigorous unification of Darwin's theory with genetics. Dubbed the "modern synthesis," it elucidated everything from the strange quirks of species found on islands to the evolving of dark coloration in British moths as the Industrial Revolution blackened trees with soot. The newly muscled-up Darwinism seemed capable of taking on every life-science mystery in sight, even one as deep as the fundamental nature of mortality.

The evolutionary ideas developed by Medawar gave unprecedented respectability to gerontology, helping to draw a new generation of scientists to the field and, in the fullness of time, to turn it into one of biomedicine's hottest areas. Some of gerontology's new talents of the 1950s and 1960s were in the same mental league as Medawar, such as Alex Comfort, a brilliant British poet-novelist-anarchist-scientist who spent years laboring on authoritative books about aging before achieving immortality by tossing off *The Joy of Sex* in a couple of weeks in 1972. (Medawar, incidentally, headed the University College London zoology department where Comfort worked.) Looking back on the still-unfolding effects of Medawar's extracurricular musings, it seems entirely possible, as gerontologist Steven Austad has suggested, that his contribution to the study of aging may someday be recognized as more important than the work that won him a Nobel Prize.

. . .

At first glance, the idea that aging cries out for an evolutionary explanation may seem counterintuitive.* After all, the world's a tough place, and just about everything in it is subject to wear and tear. Living things are clearly no exception. And while the microscopic details of biological deterioration may be a deep mystery, it would seem almost paranoid to suggest that the aging process is like the planned obsolescence of consumer goods. Our bodies may not come with extended warranties, but surely that doesn't mean Mother Nature has taken pains to equip them with subtle design flaws that ensure that they conk out after no more than about a century. Why complicate things? Why not simply think of aging as another example of the ubiquitous, essentially passive process that's responsible for everything from pot-holed streets to dulled razor blades?

Thoughtful students of Darwin, however, can easily poke holes in this commonsense argument. At the heart of evolutionary theory is a beautifully simple idea: Creatures with heritable traits that aid them in producing offspring that survive to reproductive age are more likely to pass those traits on to future generations than are peers lacking them. As a result, inherited traits that bolster reproduction tend to become more common in a species over time. Similarly, traits that diminish reproductive fitness, as Darwinians say, tend to get weeded out by the heartless process of natural selection. So why hasn't evolution produced, say, mice that live as long as we do? (Mice are generally senile by age two.) After all, a particularly robust mouse that stays frisky longer than its peers would likely generate many more offspring than they do, causing its edge to spread. Its long-lived progeny might

---

*As gerontologists are fond of pointing out, the word *aging* doesn't necessarily refer to the physical deterioration we usually associate with it—consider the aging of wine—and some of them insist on using the term *senescence* instead to refer to the process that turns us from springy youths into frail seniors. I'll stick with *aging*.

well give rise to even longer-lived offspring. Seemingly, it wouldn't take long for mice to approach our level of durability.

There's nothing specific to mice in this logic, so why don't we see heartwarming stories in the media about 90-year-old pet gerbils and 110-year-old Dalmatians? And why don't all humans live as long as France's Jeanne Calment, who was born before Custer's Last Stand yet lived long enough to release a CD partly set to rap music—she died in 1997 at 122. As E. Ray Lankester, an early authority on animal life spans, wrote in 1870, "potential longevity appears to be very nearly practically unlimited." A contemporary of Darwin (the two exchanged a number of letters), Lankester was particularly struck by evidence that many fish show no signs of deterioration with age, a mind-bending phenomenon that gerontologists now call "negligible senescence." In one notable case he cited, a 350-pound, 19-foot pike that was reportedly pulled from a German lake in 1497 bore a ring inscribed with the words "I am the fish which was first of all put into this lake by the hands of the Governor of the Universe, Frederick the Second, the 5th of October, 1230"—implying that it had inhabited the lake for 267 years after Swabia's King Frederick II released it.

Of course, it may be that Lankester was taken in by the greatest fish story ever told. But the leading expert on negligible senescence, University of Southern California gerontologist Caleb E. Finch, has compiled voluminous data from reliable, modern sources suggesting there's good reason to suspend disbelief about such cases. He cites, for example, a 152-year-old sturgeon. (Instead of looking for mementos attached to fins by governors of the universe, scientists today estimate the age of fish by examining their otoliths—pebblelike structures that accrete in their bodies like tree rings.) And for a really eye-opening example of what evolution can do for life span, consider Ming the Clam. An Icelandic quahog dredged up in 2006, and for some reason named after the Chinese dynasty, at this writing it reportedly holds the world's record for longevity among animals, 405 years.

Lest you think the blind watchmaker has a special thing for cold-blooded animals, consider the data on bowhead whales. Based on an analysis of decaying proteins in forty-eight of the big mammals' eye lenses, a University of Washington researcher estimated a few years ago that four whales harpooned by Eskimos some years earlier were more than 100 years old, and that one had reached 211. The plant world offers even more startling evidence of extreme longevity. The oldest known individual tree, a California bristlecone pine named Methuselah, sprouted more than forty-eight hundred years ago.

It's clear from all this that evolution is perfectly capable of spinning out highly durable life forms that make even the most long-lasting human body seem like a used Yugo. So why has it so rarely done so? Highlighting the curious absence of runaway life-span inflation over the eons, evolutionary biologist George C. Williams has pointed out how very strange it is that a single fertilized egg cell can miraculously transmogrify itself into a massively complex adult with trillions of cells, yet to all appearances the resulting organism is "unable to perform the much simpler task of merely maintaining what is already formed" in order to live longer. In sum, there's a profound mystery here, and it can be captured by the kind of deceptively simple questions children ask, such as, "Why did Spot, who was the same age as me, get old and die before I turned twelve?" The fact that Darwin himself conspicuously avoided grappling with such puzzles in his monumentally comprehensive tomes suggests that it can stump even the best minds.

Not every evolutionary theorist of Darwin's day realized what a Venus flytrap of a problem aging is, however, and vestiges of their dead ideas can still be found in modern encyclopedia entries. Alfred Russel Wallace, a British naturalist who independently developed a theory of evolution much like Darwin's before the more famous scientist published *On the Origin of Species* in 1859, jumped in first. In a note he jotted between 1865 and 1870, later published in a footnote by one of his correspondents, he proposed that "if individuals did not die, they would soon multiply inordinately and would interfere with each other's healthy existence." Thus, evolution "favours such races as die almost immediately after they have left successors" so that they leave enough food and other resources for their young to thrive and keep the species going.

In short, Wallace thought that evolution has equipped us all with death programs to ensure that old wastrels don't get too thick on the ground. Another early Darwinist, however, is remembered as the father of death-program theory: German biologist August Weismann. Weismann is best known for spelling out the critical distinction between germ cells (sperm and egg cells), whose regenerative powers make them essentially immortal, and the body's nonreproductive "somatic" cells, which age and die. He was also the first Darwinian thinker to dive headlong into aging theory. Echoing Wallace, he argued that "worn-out individuals" are geared by evolution to keel over in a timely way because they "are not only valueless to the species, but they are even harmful, for they take the place of those which are sound."

Weismann's view probably made a lot of sense to the young biologists of the day—junior professors at German universities generally had to wait until the occupant of an existing faculty chair died in order to get a good job. Interestingly, Weismann apparently lost his early enthusiasm for the death-program idea as he aged. Perhaps it wasn't entirely coincidental that the shift occurred after he'd been named to a chair at the University of Freiburg that he occupied for nearly forty years.

Death-program theory informed evolutionary thought about aging well into the twentieth century, and even today some theorists on gerontology's fringe push variations of it. But by the 1950s, Darwinian thinkers had generally dismissed it. They even took to ritualistically floating Weismann's theory and blasting away at it as a kind of warm-up exercise in papers on aging. That wasn't very sporting, though. Medawar had already shown it was a duck in a barrel.

• • •

Scientific theories that pass away are usually killed by conflicting data. But it turned out that Weismann's had a deeper flaw: His reasoning was circular—he blithely took aging as a given while purporting to explain why it happens. No one has highlighted the blunder better than Medawar. When Weismann asserts that worn-out individuals harm their species by shoving aside youngsters in the competition for resources, Medawar wrote, he "canters twice around the perimeter of a vicious circle. By assuming that the elders . . . are decrepit and worn out, he assumes all but a fraction of what he has set himself to prove. Nor can these dotard animals 'take the place of those which are sound' if natural selection is working," since, as Weismann surely knew but somehow overlooked, evolution favors survival of the fittest, not enfeebled seniors.

Medawar made the case for his alternative theory by imagining a lab equipped with test tubes that reproduce themselves at a constant rate. Steering clear of Weismann's circularity, he further assumed that the tubes don't age but can be destroyed by accidental breakage. As time passed, he observed, the test-tube population would consist of many youngish tubes that hadn't yet had much exposure to the risk of breakage, and progressively smaller sets of lucky older ones that had escaped clumsy handling.

Now add to the picture, he wrote, an inner cause of test-tube demise akin to getting dropped on the floor: flaws that cause tubes to spontaneously shatter at a certain age. (Such defects are like the genetic mutation that causes Huntington's disease, the inherited disorder that killed Woody Guthrie—its deadly effects don't appear until midlife.) If the shattering happens late in the tubes' normal lives, it would have a negligible impact on the population of test tubes, knocking out only a few survivors of breakage in age groups of tubes that already have made most of their reproductive contribution to the future. A lab worker whose job was to ensure that the test tubes remain plentiful might not even notice that a few old ones were spontaneously breaking—and he wouldn't have much reason to care if he did notice.

Evolution is like the lab worker—it's effectively indifferent to harmful genetic effects that drag down creatures after they've had ample opportunity to pass their genes on to future generations and have reached a time in life when diseases, predators, and accidents have left few in their age group alive. And chance—call it Murphy's Law if you want to—guarantees that mutations contributing to late-life decline will crop up. In fact, most mutations are deleterious for the same reason that randomly tweaking a well-tuned engine's bolts and belts is likely to make it splutter or die. But late-acting harmful genes aren't necessarily actively hurtful—they may simply relax defenses against wear and tear, letting creatures get trashed by random damage that they handily fended off when young. Regardless of how such genes work, though, the blind watchmaker is unable to get a purchase on them. As gerontologist Robert Arking neatly put it, the post-Medawar answer to "Why must we age?" is essentially "Why not?"

. . .

Three years after Medawar polished off Weismann's theory with his 1951 lecture at the University of London, one of America's leading biologists reincarnated it in a public talk at the University of Chicago. The speaker, Alfred Emerson, was a world authority on termites and cofounder of the "Chicago school" of ecology, whose members maintained that evolution often works to enhance the competitive edge of groups rather than of individuals. Such "group selection" seemed to explain the altruistic behavior of termites, honeybees, and other social insects whose workers are geared by evolution to act with robotic loyalty to their colonies, even forgoing reproduction for the greater good. Echoing Weismann, Emerson posited that group selection gives rise to death programs that benefit species by clearing away oldsters.

Emerson's lecture would probably have been lost to history if it hadn't been for the presence in the audience that night of a young biologist named George C. Williams. Williams, the evolutionary thinker who marveled about organisms' strange inability to maintain themselves after miraculously arising from single cells, thought Emerson was talking hogwash. Williams's wife, also trained as a biologist, attended the lecture too, and as they strolled home afterward the couple found themselves exchanging words of dismay. Williams had recently been hired to teach in Emerson's shadow at the University of Chicago, and the famous professor's talk had made him painfully aware that life sciences at the school were dominated by scholars whose views he found flat wrong.

"I remember that night very well," Williams's wife, Doris, recalled when I interviewed her in 2008. (Her husband was ill.) "The two of us were absolutely outraged at the way Emerson was talking about group selection. It was so contrary to anything we had learned or could understand. It just seemed obvious that this was not the way we wanted to view the subject. As we were walking through the night to our apartment, George turned to me and said, 'Someday I'm going to write a book to refute that lecture.'"

It was a defining moment: Williams would go on to author a series of papers and a 1966 book that convinced most biologists that the group selectionists had it wrong. Along the way he emerged as one of the century's most influential evolutionary biologists. Richard Dawkins's celebrated 1976 book, *The Selfish Gene*, which popularized the idea that evolution usually works at the level of genes rather than of whole organisms or groups, drew heavily on Williams's ideas.

True to his words that night in Chicago, Williams's initial target was the group selectionists' lingering grip on gerontology. His first major work, a 1957 paper on the evolution of aging, essentially completed the rethink Medawar had started.

A lean, bearded man with a clean-shaven upper lip, Williams during his prime bore a striking likeness to Henry David Thoreau. Besides physical similarities, the two are kindred spirits—seemingly struck from the same outcrop of flint, they represent a classic American type, the quiet iconoclast. Shy, original, stubborn, and modest to a fault, Williams stands out as one of the late twentieth century's least self-promotional giants of science. Although his ideas constantly crop up in evolutionary biology, a brief profile of him formerly offered online at Stony Brook University, where he spent much of his career, summed up his life's work in a grand total of 131 words. It didn't even mention that in 1999 the Swedish Academy of Sciences awarded him a Crafoord Prize, a Nobel Prize equivalent for disciplines not eligible for the more famous prize.

The son of a banker who was hit hard by the 1929 crash and never fully recovered, Williams grew up in the Bronx. His mother, a housewife, encouraged his boyhood interest in biology by helping him set up an aquarium and breed tropical fish. He spent endless hours musing by himself at the Bronx Zoo, one of the first places he was permitted to go alone.

After graduating from high school, he briefly served with the U.S. Army in Italy at the end of World War II, then entered the University of California at Berkeley. His first love there was paleontology, and in the summer of 1947 he spent six weeks in Arizona's Painted Desert helping one of his professors hunt fossils. But the trip's most memorable moments came when the evening talk around the campfire turned to evolution, and soon after, the subject became his intellectual passion. While finishing his doctorate at the University of California at Los Angeles, in 1954 he was hired to teach undergraduate biology classes at the University of Chicago. There he got his fateful exposure to Emerson's ideas on the evolution of aging.

Emerson "was all in favor of death," Williams later wrote, "and said that the reason we grow old and die is to make room for successors, so that they can have a chance. . . . There was absolutely no logical way you could reconcile his ideas with Darwinism, even though he claimed to be a Darwinist." In fact, "if it was biology Emerson was discussing," Williams disgustedly told himself, "I would be better off selling insurance."

Emerson was no lightweight. Among other things, his termite work helped engender the fascinating hypothesis that social-insect colonies represent "superorganisms," in which colony workers are similar to an organism's somatic cells, and queens resemble giant germ cells. What's more, group selectionists like Emerson could claim that their ideas were blessed by Darwin. In his 1871 book *The Descent of Man, and Selection in Relation to Sex,* Darwin theorized that evolution might act at the group level to engender a human tribe whose members are predisposed to "sacrifice themselves for the common good." To those disturbed by the fact that evolution entails endless, deadly competition—a process that seemed to have Tennyson's "red in tooth and claw" scribbled in blood all over it—this passage suggested a less disturbing Darwinism, one emphasizing the pluses of cooperation and self-sacrifice, albeit in the service of group competition. One member of Emerson's Chicago school even tried to wrest scientific support for his Quaker-inspired pacifism from group-selection theory.

Group selectionists tended to see "group-goodism" everywhere, and their excesses set the stage for Williams's devastating attacks. Although the debate about group selection still isn't over, Williams effectively relegated it to the role of bit player in our planet's three-billion-year

Darwinian drama. His main weapon was Occam's razor—he made it clear that attempts to explain evolved traits via group selection entail needless complexities, and that explanations that assume evolution works on individuals are more elegant and intuitively appealing. The group selectionists' fall from grace also had a lot to do with a knotty issue that often confronts tight-knit groups, such as the hippie communes of 1960s: the cheater problem.

Let's revisit the mouse world for a look at cheaters in aging. Consider a group of short-lived mice that group selection has equipped with death programs. Now imagine that one of them acquires a mutation disabling its internal time bomb, letting it live and reproduce much longer than others. Thanks to all the similarly long-lived offspring it would likely produce, its trait of longer life would spread like wildfire, aided by the fact that the longer-living mice could devour food forgone by their selfless, shorter-lived peers—a case of unleashed self-interest getting a free ride from good-for-the-group altruism. The death-programmed, altruistic mice would fairly quickly disappear, just like disenchanted 1960s commune members who moved out after getting tired of supporting the freeloaders in their midst. (The latter, of course, are drawn to communal freebies like flies to picnics.)

Pondering such logic gave Williams a particularly lively appreciation of how weird it is that so many animals age relatively fast. There just shouldn't be so many short-lived creatures around. Thus, it seemed to him that pro-aging genes didn't just randomly creep into the genome, as Medawar proposed. Rather, it seemed that some unknown force was pulling them in. After much mulling, it hit him: The genes behind aging confer a survival advantage during youth. Biologists call genes with such dual effects "pleiotropic," and because the ever-modest Williams was oblivious to the publicity advantages of clever sobriquets, his intellectual baby wound up with a name almost as endearingly ponderous as Hortense or Talmadge—it's called the "antagonistic pleiotropy theory of aging."

Despite its name, the essence of Williams's theory, like Darwin's, is simple and compelling—at least to anyone who inspects it with an open mind. (Yet, also like Darwin's, it's fraught with great hidden complexity and depth.) His key insight was that a late-acting gene that causes damage would be heavily favored by evolution if it also boosts the odds of successful reproduction early in life—even if the gene's early benefit is tiny and late damage is comparatively large. As he put it, evolution "may be said to be biased in favor of youth over old age whenever a conflict of interest arises." The result has Faustian overtones: We get the great vibrancy of youth as part of a setup that leads to our later decline and fall. Unlike Faust, however, we didn't make the deal—its terms were set when life first arose on earth more than three billion years ago.

• • •

Williams wasn't the first to invoke two-faced genes to explain aging—Medawar briefly mentioned them—but he first posited them as primary seeds of destruction. To make his case, he pictured a hypothetical gene that, before turning against us, helps calcify developing bones, strengthening them with calcium. Later in life it calcifies artery walls, inducing the familiar hardening of arteries that helps bring on heart disease and other ills. Today a number of genes are thought to fit Williams's description, such as one called p53, which helps prevent cancer's uncontrolled cell growth early in life but later appears to contribute to aging tissues' dwindling power of self-renewal.

Williams also spelled out provocative implications of his theory. For example, extreme longevity probably isn't possible unless some of the dual-acting genes that lie at the heart of aging happen to be inactive, eliminating their life-shortening effects. But if such genes were disabled, their promotion of youthful vigor would also be eliminated. Thus, he concluded, "I predict that no human being who is over a hundred years old was unusually vigorous as a young adult."

The most arresting conclusion he drew from his theory, however, was that the quest for anti-aging medicines is basically hopeless. That followed, in his view, from the fact that the body's various organs should deteriorate at roughly the same rate as a result of aging. If one bodily system, say the endocrine glands, fell apart significantly earlier than other systems, the force that evolution exerts against life-shortening genes would be focused heavily on preventing the early endocrine downfall. (Evolutionary biologist Michael Rose notes that something like this occurs when automakers recall cars to correct serious manufacturing defects—like evolution, the companies go all out to fix major glitches in newish, basically sound products.) Over time, this concentrated pushback against early endocrine aging would slow the deterioration so that it unfolded at about the same rate as aging in other parts of the body. As a result, Williams maintained, it would be impossible to isolate a few "ultimate causes" of aging. In other words, researchers who seek to slow aging are like the brave little Dutch boy rushing up to stick his finger in the dike only to find that he's become the protagonist of a tragedy and that hundreds of leaks are springing out all around him. You could almost picture Williams somberly shaking his head as he wrote,

> this conclusion banishes the "fountain of youth" to the limbo of scientific impossibilities where other human aspirations, like the perpetual motion machine . . . have already been placed by other theoretical considerations. Such conclusions are always disappointing, but they have the desirable consequence of channeling research in directions that are likely to be fruitful.

When Williams penned this blunt declaration in the late 1950s, it seemed like a principle chiseled in stone and carried down from Mount Darwin. But as we'll see, there was a totally unexpected way around it. In fact, the rate of aging in widely diverse organisms turned out to be not only amazingly plastic but also controlled in a way that has enabled scientists to slow it down with readily available interventions.

. . .

While Medawar's and Williams's overlapping theories grandly explained why we age, they said little about the molecular minutiae of aging. Then one February evening in 1977, a young British scientist, Thomas B. L. Kirkwood, emerged from his bath in a moment of high excitement and jotted down a set of ideas that extended the evolutionary theory of aging to the nitty-gritty details of bodily decay.

Born in South Africa and trained in applied math at Cambridge and Oxford, Kirkwood had landed a job a few years before at Britain's National Institute for Biological Standards and Control in London, where he was working on ways to measure proteins that make blood clot. Like Medawar, he often mused about the mystery of aging in his spare time, and like Archimedes, the Greek thinker known for bath-time eureka moments, he was given to pondering while soaking.

During his momentous immersion that night, he was mulling the idea that organisms get what they pay for when it comes to their protein building blocks. Two years earlier, U.S. and French scientists had proposed that cells can produce top-quality proteins if they spend lots of chemical energy on the process of assembling them. If they spend less energy on the process, their key components are of shoddier quality. And because a cell has only so much chemical energy, there's a trade-off: The more energy it devotes to making high-grade proteins, the less there is available for other things.

It struck Kirkwood that this trade-off may be a key factor in setting the rate of aging. That is, most of the cells in our bodies, which evolution hasn't designed to last all that long, probably devote only enough energy to molecular quality control to ensure that we don't crumble before reproducing. The exception is germ cells, which must be kept in tip-top shape to ensure that the kids are well made. Based on this insight, Kirkwood and a colleague, Robin Holliday, developed the "disposable soma" theory of aging.

The portentous name highlights the fact that, from an evolutionary perspective, our bodies are no more than disposable gene packages. That's just a clever way of encapsulating Medawar's key insight that not long after we reach the age at which our genes are usually passed to offspring, evolution effectively loses interest in our well-being, and from then on it's all downhill for us self-important, disposable gene cartons. But the disposable soma theory adds an important twist: Relatively long-lived creatures like us are geared to make relatively heavy expenditures of inner resources on cellular processes that keep protein and other molecules youthfully pristine for a long time. Short-lived animals like mice and rabbits are generally geared to invest less in such quality control, enabling them to focus more resources on developing quickly and turning out babies like mad. Roughly speaking, this means there's a kind of cosmic trade-off between sex and death—the faster a species makes babies, the faster it's likely to age.

Not all gerontologists buy this trade-off idea. For one thing, it's very difficult, and maybe even impossible, to classify units of cellular energy as spent either on maintenance or on reproduction. (Does brushing your teeth help maintain your body or make you a desirable mate, increasing your chance of reproduction?)

Still, many species' patterns of reproduction and aging are consistent with the hypothesis, and provocative support for it sprang from a landmark study in the late 1970s by evolutionary biologist Michael R. Rose.

Rose, now a professor at the University of California at Irvine, selectively bred fruit flies that were capable of reproducing at unusually old ages—ones that didn't have the right stuff to make eggs late in life weren't permitted to pass their genes to the next generation. That had the effect of exerting heavy evolutionary pressure on the insects to devote more of their inner resources to whatever it takes for a fly to keep going and stay fertile for an abnormally long time. As a result, Rose's successive generations became increasingly durable. After about a year, his artificially evolved flies lived some 10 percent longer than the ones he'd started with. Meanwhile, their fertility had dropped—they'd apparently traded early-life fecundity for more durable somas and slower aging.

Besides supporting the disposable soma theory, the study yielded an early suggestion that the speed of aging is surprisingly changeable and can even be manipulated without too much trouble in short-lived lab animals. Writing about it later, Rose marveled that "the force of natural selection easily postponed aging. I could make tiny flying Methuselahs at will."

• • •

Generating insect Methuselahs may sound like major progress toward anti-aging drugs. But Rose's remarkable study, whose results were soon confirmed by similar fly experiments led by Wayne State University's

Leo Luckinbill and Robert Arking, did little to lessen the skepticism enunciated so forcefully by Williams and continually reinforced by the anti-aging quest's endless parade of deluded dreamers and quacks. Even now, many gerontologists regard the quest as "scientific porn," according to Leonid Gavrilov, a researcher at the University of Chicago's Center on Aging.

Still, a few widely respected authorities on aging endorsed the hunt for ways to extend life span long before doing so became respectable. Peter Medawar was one of them. His interest in aging never faded, and he returned to the topic at the end of his life by penning an eloquent plea for the pursuit of therapies that lengthen healthy life span—it appeared as the final, poignant chapter of his autobiography under the title "On Living a Bit Longer." He died in 1987 after a series of devastating strokes. Although his first stroke, at age fifty-four, left him half blind, unable to use his left arm, and needing a cane and leg splint to walk, he continued his research, lectured widely, and wrote seven more books before succumbing eighteen years later. When his tribulations came up in conversation, his standard comment was, "I have a very decided preference for remaining alive."

Medawar thought highly of Williams's theory, which he termed "essentially similar" to his own. But he manifestly parted ways with the American biologist on the prospects for anti-aging breakthroughs. Looking into the future shortly before his death, Medawar foresaw the day when "people would begin to think of our allotted life as four score years and ten." Indeed, he wrote, "we can witness already the first stirrings of research of which the end result may be the prolongation of active life." The year was 1986. As usual, England's brilliant amateur gerontologist was descrying a big picture most of the pros hadn't yet caught sight of.

2

RADICALS RISE UP

IT'S IMPOSSIBLE TO say exactly when the study of aging became truly
respectable—many outsiders continued to regard the field as a speculative mess long after gerontologists began making solid progress.
But it's not hard to identify a moment in time when research on anti-
aging compounds first became an indisputably reputable pursuit. Bernard Malfroy remembers it well: Friday, September 1, 2000. That day
America's top research journal, *Science*, published a report showing
that novel antioxidant drugs he'd invented could dramatically extend
animal life span. His moment of fame began that morning, and today
he looks back on the experience with a mixture of pride, amusement,
and regret.

Malfroy, a trim, genial man with a shock of white hair, a wry smile,
and a musical accent, trained in his native France as a pharmacologist,
then launched his biotech career in 1985 as a researcher at Genen-
tech, the industry leader in South San Francisco. In 1989 he moved to
Alkermes, a biotech start-up in Cambridge, Massachusetts. The early
1990s, however, was a time of entrepreneurial exuberance in biotech,
and in 1991 Malfroy founded his own company, Eukarion. Installed
in one of the low-rent commercial parks along Boston's Route 128,
the start-up focused on developing antioxidants to neutralize free

radicals—the highly reactive molecules that are often portrayed as a slow-acting poison that drives aging.

The idea that taking antioxidants might make us age gracefully and live longer has probably been embraced by more people than any other concept from gerontology. Its magnetism emanates from the attractive hypothesis behind it—the free radical theory of aging—which collapses the dizzying complexity of aging into a simple problem that we can readily address. The theory holds that free radicals basically rust out our cells, and it follows that we can counter creeping body rot with radical-neutralizing antioxidants. At least, that's what dietary-supplement and food makers want us to believe. And we buy it—lots of it. Annual sales of antioxidant supplements, such as vitamin C and blueberry extracts, exceed $3 billion. You can even get ice cream and chewing gum spiked with antioxidants.

Eukarion's compounds were no ordinary antioxidants, though. Designed to mimic potent enzymes that naturally exist in cells to quash free radicals, they were superheroes of the antioxidant world, and Malfroy dreamed of turning them into lucrative, FDA-approved drugs for diseases of aging. The villains they were set to vanquish turned up left and right during the 1990s—free radicals were implicated in just about every degenerative illness, from Alzheimer's to heart disease. Then came the electrifying *Science* paper.

The study's subjects were comma-sized worms, officially known as *Caenorhabditis elegans*, from Greek and Latin words for "elegant new rod." They're members of a diverse group of soil dwellers, called nematodes, and they turn up especially often on the frontier of science. The worms are readily propagated in the lab and simple to maintain— thousands can grow in a single petri dish. They're transparent, so you can easily witness their rich inner lives. Of special interest to gerontologists, they have an average life span of two to three weeks, so you don't have to wait long to see whether possible anti-aging drugs have an effect on them. And *C. elegans* researchers are famously collegial— worm people have a reputation for being nearly as nice to work with as nematodes themselves.

Gene mutations that slow aging first came to light in studies with

C. elegans in the late 1980s. (More on that later.) But before 2000, no drug had been shown to extend the worms' lives, or those of any other species, in the quick-get-me-some-of-that-stuff way that Eukarion's did. In fourteen experiments with different sets of worms, one of its compounds, EUK-134, consistently boosted the animals' average life span by nearly half. The appearance of the spectacular data in America's number-one science journal bestowed unprecedented prestige on the anti-aging quest.

It didn't hurt that the study was instigated and overseen at the Buck Institute for Age Research in Novato, California. The institute is perched on a wooded mountain twenty-five miles north of San Francisco, housed in an elegant temple of science designed by architect I. M. Pei. Set up in 1999 with a grant from the estate of philanthropists Leonard and Beryl Buck, it's the only independent U.S. research center focused solely on aging and age-related diseases. The *Science* paper based on Eukarion's compounds was the center's first major claim to fame. In a press release, the study's principal author, Buck Institute gerontologist Simon Melov, summed up the data as "the first real indication we have had that aging is a condition that can be treated through appropriate drug therapy." Further studies with Eukarion's drugs, he continued, "will allow us to answer whether or not we have to reconsider aging as an inevitability."

The media loved it, especially British headline writers, who have a constitutional weakness for puns. WORM HAS TURNED IN ANTI-AGEING FIGHT, chuckled one UK newspaper. Another UK headline ventured that WORMS COULD HELP US ALL TO WRIGGLE OUT OF OLD AGE. Others simply stuck to old-fashioned, straight-ahead hype, such as, HUMBLE WORM HOLDS SECRET OF ETERNAL YOUTH. The story also got wide play in North America, and Malfroy found himself on a Canadian radio talk show with a fellow "expert" who declared that immortals were already walking the earth. After Malfroy pointed out that that was ludicrous, there was an extended moment of silence on the airwaves.

Malfroy and other Eukarion scientists took pains to disabuse reporters of the idea that the biotech planned to develop anti-aging drugs. Their goal had always been to treat diseases of aging, not to try tampering with the aging process.* In an interview with the British newspaper the *Guardian*, Susan Doctrow, Eukarion's vice president of research, noted that "we're not going to test our compounds for their effects on ageing. But if the effect of treating diseases of old age is to extend life, everyone's going to be happy."

Never before had hopes for turning basic insights from gerontology into breakthroughs seemed so realistic. Nor had the idea that free radicals are time's main wrecking balls, first proposed in the 1950s, ever seemed so right.

Of course, only a small fraction of drugs that show promise in lower animals like worms prove safe and effective in people. But worm aging is surprisingly similar to human senescence. As nematodes get old, their muscles weaken, their motor activity declines, they lose their appetites, they take on a rough and lumpy appearance—they even get constipated. Thus, it wasn't unreasonable to hope that medical science had just taken a giant step toward life-span extension. Even if Eukarion's drugs failed to pass muster in clinical trials, it arguably would be only a matter of time—and probably not a very long time at that—before even better antioxidants worked anti-aging magic in people.

Unfortunately, things weren't that simple.

. . .

The free radical theory of aging is more than fifty years old, and the man who thought it up was almost forty when it occurred to him. After adding those numbers together, you wouldn't expect to walk down a hall at a medical research center today and spot his name on one of the doors. But when I sought him out in mid-2008, I found

---

*Contrary to the assertions of some who would "cure" it with their favorite elixirs, aging isn't a disease. Diseases afflict subsets of the population; aging hits everybody. And it would be counterintuitive, even semantically reckless, to assert that a person at fifty is sick because he's less adept at learning to play a musical instrument or at turning cartwheels than he was at twenty—Michael Jordan didn't retire from the NBA because he was ill. Of course, there's a close link between aging and sickness—some scientists define aging as a bodily process that increases vulnerability to disease.

Denham Harman, at ninety-two, still working at the University of Nebraska Medical Center in Omaha, where he had spent the past half-century. Though somewhat frail and hard of hearing, he was professorial and precise, casually natty in a white shirt, gray slacks, and maroon tie, and continuing to make history—or at least to nail down his place in it; when I dropped in, he was working on a paper about key milestones in the study of free radicals and aging. He insisted that I take the most comfortable chair in his office, and after I reluctantly accepted, it took me a while to stop marveling about the fact that I was interviewing an Authentic Historic Figure—it seemed a little like opening a door into the medical arm of the Twilight Zone and finding Louis Pasteur scribbling away on a germ-theory paper.

A native of San Francisco, Harman came of age during the Depression in a family whose fortunes rose and fell with the stuttering economy. During his high school years, his father, a gentlemanly, London-born accountant who had come to the United States in his early twenties, sometimes had trouble getting work, and for a while it seemed that college would be financially out of reach for Denham. Then a chance meeting at a Berkeley tennis club between his father and a fellow English expatriate, who happened to be the director of Shell Oil's research arm, helped him land a job as a lab technician at the oil company's research center in Emeryville, California. It paid sixty-five dollars a month, Harman recalled with characteristic exactitude, and after his years of peddling newspapers on street corners to help make ends meet, "that was a lot of money." (One of his most vivid early memories was the time he gave in to the rare, guilty pleasure of spending some of his hard-earned newspaper money on a root beer on a sweltering summer day—a vignette right out of Norman Rockwell. He could still taste the cold, sweet, nose-prickling fizz nearly eighty years later.) Urged by his boss at Shell to pursue a career in chemistry, he soon saved up enough to enter the University of California at Berkeley. But he held on to his job at Shell for more than a decade while going to school, learning the ropes as an industrial chemist as he earned his Ph.D.

While America's top physicists raced to build the atomic bomb during World War II, chemists at major universities like Berkeley found themselves swept up in military research that had the same damn-the-torpedoes urgency as the Manhattan Project. One of Berkeley's promising young chemists, a classmate of Harman's, lost his life in this behind-the-scenes war of the chem labs: "He was working with phosgene [a chemical-warfare agent], and one day he broke a flask, got a whiff, and just lay down," Harman said. "He knew he was going to die, and he did."

For his part, Harman took part in pioneering research during the 1940s on free radicals, which were increasingly recognized as key intermediaries in chemical reactions—the unstable molecules often flicker into existence as reactions unfold and speed up the chemical changes taking place. At Shell, he and colleagues sought to harness free radicals to help synthesize products such as pesticides and polymers. Harman racked up three dozen patents during his years at the company, mostly based on his free radical research.

By the end of the war, he seemed well on his way to becoming a distinguished industrial chemist. Then one evening in December 1945 his wife handed him a *Ladies' Home Journal* article written by prominent *New York Times* science writer William L. Laurence, the official journalist of the Manhattan Project, titled "Tomorrow You May Be Younger." The subject was a Russian scientist who claimed to have invented an anti-aging compound. In a memorably ghoulish twist, he reportedly extracted his elixir from the blood of horses injected with tissues from healthy, young humans who had died by accident. It was yet another example of the sensational silliness that drives mainstream gerontologists crazy, but it sparked Harman's interest in aging and helped convince him, at thirty-three, to change directions and go to medical school.

Five years later, in 1954, Dr. Harman, M.D.-Ph.D., was hired as a research associate at Berkeley's Donner Laboratory of Biophysics and Medical Physics. Its director, John Lawrence, known as the father of nuclear medicine, was pioneering the use of radioactive isotopes to treat cancer. Harman's formal duties were light: He was required to spend a few hours each week examining cancer patients, which gave him time to launch a research project of his own devising. Going for broke, he decided to seek the fundamental cause of aging. He figured his unusual background in chemistry and medicine might let him see things that had eluded others.

"Everything dies," he told me, "so I thought there had to be some common, basic cause. I sat at my desk for four solid months without getting anywhere. It was damn frustrating. Then, when I was just about ready to chuck the whole business, the phrase 'free radicals' crossed my mind one morning in November. I thought, 'My god, have I got the answer?'"

It didn't take him long to convince himself that he did. Thanks to his research at Shell, he was acutely aware of free radicals' power to break the chemical bonds that hold atoms together in molecules. Thus, he could readily picture how the radicals might wreak havoc in living cells.

I myself can't hear the term free radical without picturing Abbie Hoffman attempting to levitate the Pentagon with psychic energy, but in chemistry it denotes a molecule with an unpaired electron. (Electrons, the negatively charged particles that buzz around the nuclei of atoms, are usually found in pairs, and things tend to get messy when only one of a pair is present.) To eliminate its electron deficiency, a free radical typically grabs an electron from a nearby molecule. Because electrons form bonds between atoms, the molecular victim of this act of violence can be severely deformed or even broken up by its loss of a key structural component. Worse, lacking one of its electrons, it, or one of its remnants, becomes a free radical that's likely to steal an electron from another innocent bystander. Within microseconds, a chain reaction can erupt that creates a widening circle of damage to a cell's proteins, lipid molecules, and DNA.

The most important free radicals in biology consist of oxygen bonded to hydrogen and other elements—scientists often refer to them as reactive oxygen species, or ROS, and to the harmful effect of run-amok mobs of free radicals as oxidative stress. The most dangerous one is the hydroxyl radical, an oxygen-hydrogen combination with a particularly ferocious appetite for electrons. Another key member of the ROS family is hydrogen peroxide, found in antiseptics and bleach. Although technically not a free radical, hydrogen peroxide is continually formed in cells from a free radical, called superoxide (a radicalized oxygen molecule), that's released by the energy-producing oxidation of sugar. In fact, every breath we take engenders tiny puffs of the stuff of bottle blondes in our cells. That's not usually a problem—peroxide by itself isn't very reactive—but in the presence of metals like iron it breaks down to form the highly reactive hydroxyl radical. Not surprisingly, our cells handle iron atoms very gingerly, normally keeping them under wraps inside large protein "complexes," such as one at the center of hemoglobin, the protein that carries oxygen in red blood cells.

Antioxidants are chemicals that can yield up electrons to free radicals without becoming greedy electron thieves themselves. Chemists call them chain-breakers, because their electron donations halt free radical chain reactions. Curiously, vitamin C actually turns into a free radical when it plays chain-breaker. But the arrangement of electrons in radicalized vitamin C molecules keeps them relatively stable, making them much less reactive than most free radicals. Vitamin E molecules act the same way. But they often don't stay in free radical state long because vitamin C steps up to restore their lost electrons. This teamwork literally covers the waterfront inside our bodies. Because it's fat-soluble, E plays chain-breaker in lipid-rich places, such as cell membranes and cholesterol molecules, while water-soluble C halts chain reactions in the surrounding watery realms.

Many of these free radical basics were known, or at least suspected, when Harman sat down to solve the mystery of aging. Six months before his aha moment, a team led by Rebeca Gerschman, a University of Rochester researcher investigating radiation injury under the auspices of the federal Atomic Energy Commission, had proposed that tissue damage from exposure to X-rays, or breathing pure oxygen (it's much deadlier than tobacco smoke), stems from the release of free radicals in cells. A second prominent 1954 report on free radicals caught Harman's attention: Researchers led by cell biologist Barry Commoner, who later won fame as an environmentalist, demonstrated that free radicals could be detected in cells using a technology akin to that in today's MRI scanners. Before Commoner's finding, the molecules' fleeting existence had thwarted attempts to prove that they really do exist in living tissues.

But during the 1950s most scientists believed molecules as ephemeral as free radicals couldn't possibly play important roles in biology. Thus, Harman's conclusion that they were behind aging wasn't at all obvious. In 1956 Harman first presented the theory to a wide audience in a succinct, two-page report in the *Journal of Gerontology*. (A version of it had appeared the previous year in his lab's newsletter.) The paper, in retrospect, was amazingly prescient. Harman posited that the hydroxyl radical, generated in the course of normal metabolism, is probably a primary agent of our destruction; that DNA damage from free radicals might cause gene mutations that lead to cancer; that heart disease may result from free radical injury to cells of the circulatory system; and that free radical damage might impair cells' "functional efficiency" and capacity for self-renewal. Each of these ideas is now supported by a mountain of data.

Not one for coyly skirting real-world implications, Harman boldly stated in his 1956 paper that his theory was "suggestive of chemical means of prolonging effective life." Mouse experiments were already under way, he added, to test whether compounds that counter free radicals might ward off cancer and perhaps even slow aging.

Like many ideas ahead of their time, Harman's fell stillborn from the press, to use philosopher David Hume's rueful phrase about his own magnum opus. At Donner Lab, Harman recalled, friendly colleagues told him that his theory was intriguing but too simple to explain aging. Elsewhere, "it was either ignored or ridiculed." Harman began worrying that he wouldn't be able to make a living solely doing research. Marked by the Depression, he was loath to waste anything, including his years of training to be a doctor. Soon after proposing his free radical theory, he cut back on his hours at Donner to complete residency requirements for practicing medicine. Then, after two of his physician acquaintances at the lab threw in the towel and returned to doctoring, he too dropped out of Donner's illustrious rat race and in 1958 joined the faculty at the Nebraska medical center.

The morning I visited him happened to be fifty years to the day since he had set out for Omaha. "I hated to leave San Francisco," he reflected, sitting beneath a large painting of wind-whipped Pacific waves hanging over his desk—the most striking personal effect in his office. Still, he apparently looked back on the beginning of his new life as a treasured memory. As I listened to him reminisce, it seemed like only yesterday that he was driving toward the rising sun with his wife, his three sons, and the family's pet hamster.

• • •

It took more than a decade for Harman's theory to get traction after its 1956 debut. The boost that started it rolling into biology's foreground was provided by a Duke University biochemist named Irwin Fridovich and one of his graduate students, Joe M. McCord. In 1968 they isolated an intriguing blue-green protein, later dubbed superoxide dismutase, or SOD, from red blood cells of cattle. They knew it was an enzyme because, like all such catalytic proteins, it greatly accelerated a specific chemical reaction—in SOD's case, the conversion of superoxide free radicals into hydrogen peroxide and oxygen. But the full import of the discovery didn't become clear until they and other researchers investigated SOD's role in living cells. Because the enzyme could be readily detected, it served as a kind of divining rod for will-o'-the-wisp free radicals. "We were like children with a new toy or like a craftsman with a new tool," Fridovich wrote years later.

One of SOD's most striking properties, they found, was its ability to eliminate superoxide at an almost impossibly fast rate, an indication that it was designed by evolution to handle surprisingly large amounts of the free radical in cells. Its ubiquity stood out too. SOD variants were found everywhere scientists looked, from bacteria to human cells. Gradually it became clear that free radicals are constantly churned out by cells' oxygen-based energy systems, and that SOD is a key defense against their harmful effects.

But it also became apparent that superoxide isn't always instantly disposed of, supporting Harman's hypothesis that free radical damage can happen. For one thing, researchers discovered that white blood cells, like tiny dragons, spit superoxide at invading bacteria. And although superoxide isn't terribly reactive as ROS go, it can team with hydrogen peroxide to produce the highly destructive hydroxyl radical. All in all it looked as if Harman had been right to finger free radicals as prime suspects in the great biochemical murder mystery of life.

Ironically, just as Harman's theory was gaining credibility, he himself was having doubts about it. Between the late 1950s and late 1960s, he had put his theory to the acid test by feeding antioxidants to different strains of mice to see whether they lived longer than control animals. At first blush the results seemed promising. Some, but not all, of the compounds increased the "half-survival time"—the age at which half of a group of mice were still alive—by as much as a third in strains of short-lived, cancer-prone mice. However, it was entirely possible that the antioxidants had only delayed the appearance of tumors in the mice rather than retarded their aging process.

Mindful of that possibility, Harman next carried out similar experiments with male mice of the "LAF1" strain, which have relatively low tumor incidence. The results, reported in 1968, showed that 2-mercaptoethylamine, an antioxidant sometimes used as a treatment for radiation sickness, lengthened the rodents' average life span by nearly 30 percent. Meanwhile, several other researchers were following in his footsteps. In 1971, Britain's Alex Comfort reported that an antioxidant called ethoxyquin significantly boosted life span in one of the same strains of short-lived, cancer-prone mice Harman had used. Comfort's colleague at University College London, Peter Medawar, the eminent immunologist who had shed light on the evolution of aging, was impressed by the mouse studies, and during his later years he took large daily doses of antioxidant vitamins C and E in hopes of extending his own life.

But the rodent studies still didn't make a strong case for Harman's theory. The data were mixed—scientists' euphemism for "we really can't tell what the hell's going on." Some kinds of mice didn't live longer on any of the antioxidants that were tried. Some antioxidants that seemed to work did so only in certain mouse strains. A few of the compounds were downright toxic to the rodents at the applied doses. Further, the antioxidants that worked best in mice posed toxicity risks to humans and other animals. Ethoxyquin, for example, has been linked to kidney and liver toxicity in several species.

What most troubled Harman, however, was the fact that although antioxidants sometimes raised life expectancy, or average life span, they never increased maximum life span. That had the look of a killer issue. Harman acknowledged in a 2003 interview, "I had expected both of them, mean [average] and maximum life spans, to be increased [by antioxidants given to mice]. So I came to a halt for quite a while" to ponder two big questions: "Was the theory wrong? Did I miss something or what?"

To understand his crisis of confidence, consider the effect of installing a waterworks in a rural town in Africa that has long relied on a polluted community well. It's likely that over time the life expectancy (reminder: that's average life span) of the town's inhabitants would rise, largely because of lower infant mortality from water-borne infections. (Such public-health measures were largely responsible for a rise in U.S. life expectancy by nearly thirty years between 1900 and 2000, from about age forty-seven to about age seventy-seven.) But drinking clean water wouldn't make the people actually age slower and live longer than healthy people typically do in other places.

In contrast, if half of the people in the town were still sprightly at one hundred after taking long-term, daily doses of, say, a novel antioxidant, there would be little doubt that the compound had profoundly altered their aging process. There's no other way their maximum life span could have risen so dramatically. (A commonly used proxy for maximum life span, by the way, is the average life span of the longest-lived 10 percent of a study group.) Thus, maximum life span is a much better indicator of anti-aging effects than life expectancy is.

So here's the possibility that troubled Harman: Antioxidants' consistent failure to extend rodents' maximum life span suggested that the benefits they sometimes conferred may have resulted from amelioration of diseases, as providing clean water does, rather than slowed aging. This kind of ambiguity crops up surprisingly often in gerontology. One reason is that supposedly normal rodents used in studies on aging are often disease-prone and short-lived. Many strains of lab rodents were drafted for research precisely because they're predisposed to life-shortening disorders of medical interest. And widely used strains of lab mice, like livestock, have been bred to mature fast and reproduce rapidly—in keeping with the idea that there's a trade-off between fertility and longevity, they're very fertile but may well age faster than their wild ancestors. That may make them ideal for studies on the fallout from modern lifestyles.* But it can be hard to say whether a drug that boosts longevity in lab mice has slowed normal aging or has merely ameliorated life-shortening disorders peculiar to hothouse animals with idiosyncratic genomes.

Although taken aback by the murky antioxidant data, Harman wasn't about to abandon his baby. In 1972, he proposed a second big idea about free radicals that both explained the troubling antioxidant results and pointed to a fertile new frontier in free radical research. His updated theory stressed that free radicals mainly arise inside mitochondria—power plants inside cells where oxidation of sugar releases energy. Mitochondria are crammed with delicate machinery that's highly vulnerable to free radical damage. Each mitochondrion has its own set of thirty-seven genes separate from those in the cell nucleus, as well as lipid membranes and complex sets of energy-generating proteins—mitochondria are actually remnants of ancient bacteria that took up residence nearly two billion years ago as symbiotic organisms inside larger cells.

Cells' native antioxidants, such as SOD, neutralize most of the free radicals spun off by their power plants. But some radicals elude the defenses and act like wrenches hurled into spinning dynamos. As a result, Harman suggested, mitochondria are the key sites of free radical injury behind the aging process. He compared them to "biologic clocks"—when the damage reaches a critical mass, they burn out in various key organs, and death ensues. It's probable, he added, that membranes surrounding mitochondria, which fence them off from other parts of cells, block out antioxidants ingested in food or drugs, preventing them from reaching the key sites of free radical injury.

---

*Contrary to the assertions of some who would "cure" it with their favorite elixirs, aging isn't a disease. Diseases afflict subsets of the population; aging hits everybody. And it would be counterintuitive, even semantically reckless, to assert that a person at fifty is sick because he's less adept at learning to play a musical instrument or at turning cartwheels than he was at twenty—Michael Jordan didn't retire from the NBA because he was ill. Of course, there's a close link between aging and sickness—some scientists define aging as a bodily process that increases vulnerability to disease.

That's why feeding the compounds to mice didn't extend their maximum life span.

But taking antioxidants might confer significant health benefits, he theorized, by blocking damage that occurs outside mitochondria. Research by other scientists has elucidated how such damage occurs. Free radicals outside mitochondria, for instance, have been shown to help foster "cross-linking" reactions in which sugar molecules react with proteins to form "advanced glycation end products," or AGEs, in which the protein molecules are basically glued together in an irreversible way that stiffens and degrades them. AGEs engender yet more radicals, accelerating the sugary decay of proteins—this may well be one of the worst of the bad things that happen to us as we age. (Picture a bunch of stiff, old stuck-together rubber bands—those are your cross-linked proteins at age eighty.) Cross-links in collagen alone, an abundant protein in skin, cartilage, arteries, and tendons, are thought to underlie everything from high blood pressure to wrinkles.

Harman's updated theory was later honed by other scientists, who postulated that damaged mitochondria, like aging engines belching smoke, tend to spew more free radicals, which causes yet more mitochondrial damage—killer positive feedback, in other words. According to a popular variant of this idea, the degradation of mitochondrial DNA is the main driver of a self-accelerating "vicious cycle" at the heart of aging.

During the 1980s and after, biologists found more and more to like about Harman's matured brainchild. In 1988, Bruce Ames, a prominent biochemist at the University of California at Berkeley, added weight to the theory by reporting that signs of free radical damage in rats' mitochondrial DNA increase as they age. A few years later Takayuki Ozawa and colleagues at the University of Nagoya in Japan showed that mitochondrial DNA damage rises exponentially with age in humans, just as mortality risk does. In 1992, Earl Stadtman at the National Institutes of Health marshaled evidence indicating that nearly half of the protein in elderly bodies is scarred by free radical damage. Other researchers reported that genetically altered fruit flies with bolstered antioxidant enzymes live longer than normal flies.

By the mid-1990s, proponents of the free radical theory could boast more experimental support than had ever been mustered in favor of a big idea about aging. And though Warren Buffett's billions were all very nice, it was arguable that the real sage of Omaha was Denham Harman.

. . .

If any biotech company could claim to be on top of the rising wave of support for the free radical theory, it was Eukarion. Other biotechs were riding the wave: In 1994, Fridovich, the discoverer of SOD, cofounded Aeolus Pharmaceuticals to capitalize on his research. Garland Marshall, a biochemist at Washington University in St. Louis, formed MetaPhore Pharmaceuticals in 1998 to work on another set of SOD-like drugs. And in 1999, Berkeley's Bruce Ames cofounded a company called Juvenon to sell dietary supplements designed to help protect mitochondria from free radicals. But Malfroy's company had a head start, and a number of animal studies had suggested that its compounds might be able to ameliorate neurological diseases, heart attacks, and other old-age killers.

The idea of developing drugs that mimic the body's own extremely potent free radical defenses dated from the discovery of SOD in 1968. For years after, researchers hoped to turn SOD itself into a medicine. The enzyme promised to do great things—as a free radical remover, SOD is to familiar antioxidants like vitamin C as an industrial vacuum cleaner is to a whisk broom. But the idea didn't pan out. Given as medicine, SOD was eliminated too fast in the body, provoked immune reactions, and was formidably expensive.

When Malfroy founded Eukarion in 1991, researchers had mostly given up on SOD and turned to synthetic compounds that mimic its catalytic action. Unlike SOD, a large, fragile protein molecule, the new drug candidates were relatively durable, small molecules that were more likely to reach sites of free radical damage. The seed idea for Eukarion's SOD mimetics sprang from regular bridge games Malfroy played with a software executive who lived in his neighborhood near Boston. One day the neighbor introduced him to a computer program for chemists whose training examples featured an industrial reagent that looked a bit like SOD. (To be precise, it resembled one of three kinds of SOD, called SOD2, that specializes in defending mitochondria.) Within a few months Malfroy had established that a version of the molecule showed promising SOD-like activity in the test tube.

He and Doctrow, who together led Eukarion's research, later discovered that their prototype drug, EUK-8, is a dual-action antioxidant—it mimics the antioxidant functions of both SOD and catalase. That was a major stroke of luck. Drugs that boost only SOD, which turns superoxide into hydrogen peroxide, would potentially backfire by generating an excess of peroxide in cells. But with EUK-8, Eukarion could avoid the risk, because the compound would boost the activities of both catalase and SOD, and the revved-up catalase would then convert hydrogen peroxide to water and oxygen as fast as the boosted SOD churned out peroxide. One cold, dark day late in December 1993, Malfroy rigged up a simple but striking test of EUK-8's power: When he added a sample of the compound to a test tube full of hydrogen peroxide, bubbles of oxygen instantly fizzed up—a sign of the same kind of free radical neutralizing that happens in cells. He dubbed it the "champagne experiment."

A CEO who preferred working in the lab to pursuing potential investors, Malfroy kept Eukarion small, making do with modest backing from wealthy "angel investors" and research grants from the National Institutes of Health rather than trying to turn the company into a glitzy, go-go biotech bankrolled by venture capitalists. The strategy arguably worked too well. Joining forces with dozens of academic collaborators, Eukarion consistently punched above its weight in the medical literature. In the late 1990s, its promising preclinical results inspired Glaxo Wellcome (now GlaxoSmithKline) to fund exploratory research on its compounds. But the collaboration failed to yield compounds that the big drug company deemed ready for clinical testing, and Glaxo walked away after two years, leaving Eukarion stuck in research-boutique mode with little means of support. Scraping by on a shoestring budget with a staff of ten, it couldn't do extensive animal tests, which can cost millions of dollars, much less clinical trials, which can cost hundreds of millions. By the end of the decade, Malfroy found himself grappling with a biotech version of a Catch-22: To take the company to the next level, he needed a major cash infusion to pay for big-ticket studies. But to convince investors that Eukarion warranted a hefty cash infusion after nearly a decade without putting a drug into the clinic, he needed data from the kind of studies that his start-up had never been able to afford. "What allowed us to survive all those years was having a very lean operation," Malfroy said. "But that also prevented us from succeeding. We realized that we had boxed ourselves in."

Then the riveting worm study in *Science* appeared, seeming to hand him a solution to his Catch-22 problem. After it was published he got a call from a venture capitalist eager to discuss a major financing. Estée Lauder, the cosmetics giant, wanted to add Eukarion antioxidants to products that might slow skin aging. A Swiss biotech company, Modex Therapeutics, licensed one of Eukarion's compounds and began testing it for preventing and treating skin burns from radiation therapy administered to cancer patients. Meanwhile, academic researchers launched a new wave of studies with its compounds.

In the end, though, it was Malfroy's hopes, not his fiscal conundrum, that were crushed. The bursting of the dot-com bubble after 2000 caused massive collateral damage in other high-tech niches, and investor interest in biotechs like Eukarion shriveled a few weeks after Malfroy began talking with VCs about a major financing. Then the University of Southern California's Rajindar Sohal, a big name in free radical research, reported in mid-2002 that Eukarion's compounds failed to extend life span in houseflies. Six months later, researchers at University College London published a study indicating that EUK-8 actually shortened rather than lengthened nematodes' life spans. The discouraging data didn't prove that the earlier worm study in *Science* was wrong—differences in the methods used in the different labs, such as varying ways of administering Eukarion's compounds, might have caused the different outcomes. But the back-to-back failures to confirm the earlier results dimmed Eukarion's glow.

Still, Eukarion's compounds continued to rack up encouraging data in studies based on animal models of human disease and aging. Few of the reports made the news, but some were arguably more provocative than the famous life-extension study. A 2003 mouse study, for instance, offered hope for quelling the senior-moments epidemic that's sweeping the world as baby boomers turn gray: It showed that chronic doses of Eukarion's experimental medicines almost completely reversed learning and memory deficits that normally afflict middle-aged mice. Overseen by University of Southern California neuroscientist Michel Baudry, who had helped Malfroy start Eukarion, the study showed that the mental sharpening afforded by the drugs closely tracked a reduction of oxidative stress in the rodents' brains. Such stress is thought to be a major contributor to Alzheimer's disease, brain damage from strokes, Parkinson's disease, and many other brain disorders, as well as the normal decline of mental acuity with age.

But the heartening studies didn't stop Eukarion from reaching a fiscal dead end. Ironically, the crunch came right after Estée Lauder licensed a Eukarion compound for use in skin-care products. One Friday in early April 2002, Malfroy traveled to the cosmetics company's New York City headquarters for a meeting with its chairman, Leonard Lauder, son of the founder. Since Lauder's scientists were enthused about Eukarion's compounds, Malfroy recalled, "I had high hopes" the wealthy executive might personally invest in the struggling biotech. "But nothing happened at the meeting" to suggest he would—and he didn't.

That evening Malfroy and his wife attended a Carnegie Hall concert commemorating cellist Mstislav Rostropovich's seventy-fifth birthday. The performance had a valedictory aura—the old master, who died in 2007, had made his American debut in 1956 at Carnegie. Malfroy found himself especially attuned to the occasion's farewell feeling: After walking away empty-handed from his audience with Lauder, "I decided that day the only possibility for Eukarion was to be acquired," he said.

After winding down Eukarion's operations, he sold it in late 2004 to one of its research partners, Proteome Systems of Sydney, Australia. The purchase price, according to the *Boston Business Journal*, was $1 million in stock and "a modest cash payment." The deal also included possible future payments of more than $20 million that were contingent on achieving certain milestones with Eukarion's drugs. Deeply frustrated but still hopeful, Malfroy and Doctrow stayed on at Proteome to lead research on their compounds. In 2005, a collaboration Doctrow had arranged with a colleague at the Medical College of Wisconsin led to a grant from the NIH, which selected the antioxidants to test as treatments for people exposed to radiation in terrorist attacks and industrial accidents. Three years later, however, Proteome terminated its drug-development work, including research on Eukarion's compounds, in order to focus on diagnostics. Refusing to give up, both Doctrow, now at Boston University, and Malfroy, who founded a new company called MindSet-Rx, have continued to pursue development of the experimental drugs as therapies for people exposed to radiation and for patients with rare neurological diseases.

. . .

Eukarion wasn't the only company that had trouble capitalizing on free radical science. In 2005, MetaPhore's main drug candidate failed in two clinical trials, prompting the company to merge with another struggling biotech that later folded. After more than a decade of work, Aeolus, the biotech cofounded by Fridovich, had nearly run out of money, and in 2007 it announced that it would postpone trials on its primary drug candidate, conserving capital while the National Institutes of Health sponsored research on the drug as a treatment for mustard gas exposure.

By mid-decade it was clear that breakthrough antioxidant drugs, once gerontology's most promising commercial prospect, weren't imminent after all. Management missteps, bad luck, and the insanely high risk of drug development were partly to blame. But there was also a deeper problem: A new chapter in free radical science had opened, and one of its surprising twists is that antioxidants don't always come off as good guys.

Early hints of the revisionism arose in a long-running debate about vitamin C. The controversy ignited in 1970 when two-time Nobel laureate chemist Linus Pauling claimed in a popular book that downing gobs of the vitamin could prevent or cure the common cold. Pauling advised taking more than 2 grams a day, more than twenty times the generally recommended daily dose (the current U.S. "reference daily intake" is 75 milligrams for women and 90 milligrams for men). He himself reportedly took up to 40 grams a day. In a series of books and technical reports after 1970, he argued that megadoses of vitamin C could ward off cancer and heart disease, boost mental alertness, and maybe extend a person's life by decades. Pauling, who died from prostate cancer in 1994 at age ninety-three, believed that our mammalian ancestors got massive amounts of vitamin C in their plant-rich diets, and that we're geared by evolution to need far more of it for optimal health than we get from modern diets.

No one disputes the critical importance of vitamin C as a dietary component. Identified in 1932, it's a remarkably versatile nutrient—and not only as an antioxidant. Vitamin C is essential for making collagen. We also need it to make neurotransmitters that carry signals between nerve cells. It facilitates absorption of dietary iron, and without it anemia sets in. No wonder most animals can synthesize it in their bodies, hence aren't reliant on dietary sources. The few oddballs that lost the ability to make vitamin C at some point in the evolutionary past include humans and other primates, guinea pigs, and fruit-eating bats. In Pauling's view, the deletion of our ancestors' ability to synthesize vitamin C was a very bad move by Mother Nature, making us vulnerable to a life-shortening deficit of the vitamin as we moved away from plant-heavy diets.

His view has some merit, as famously shown by the prevalence of scurvy on British navy ships before their sailors were issued citrus rations after 1795, hence their nickname, "Limeys." (Scurvy, caused by getting less than about 10 milligrams of vitamin C a day, induces extreme fatigue, bleeding gums, swollen limbs, and, ultimately, heart failure.) But as Pauling's critics have made abundantly clear, there's no compelling clinical evidence that megadoses of vitamin C offer significant benefits. Some of his logic seems iffy too. For instance, evolution appears to have taken pains to prevent us from getting too much vitamin C, as if it were potentially harmful. In fact, when we ingest more than 60 milligrams a day, we begin excreting it in our urine, and the daily dose we can absorb tops out at about 400 milligrams. This is odd: After we lost the ability to make vitamin C, why wouldn't evolution have geared us to sock it away in our tissues for a rainy day, especially if it's as amazingly beneficial as Pauling insisted?

Free radical experts have suggested a shocking answer to this question: Vitamin C can abet free radical damage as well as prevent it. For instance, it can combine with iron and hydrogen peroxide to release the dynamitelike hydroxyl radical. That probably accounts for data suggesting that ingesting large doses of iron plus vitamin C induces free radical injury in the intestines, potentially increasing the risk of ulcers, cancer, and inflammatory bowel disease. Other troubling research indicates that large doses of vitamin C can attenuate the body's natural antioxidant defenses.

This last downside looms large, for it probably isn't specific to vitamin C. There's evidence that taking large doses of potent antioxidants can discombobulate intricate mechanisms in cells whose function is to counteract oxidative stress. The case for this distressing conclusion rests partly on the fact that, as strange as it may seem, free radicals serve as chemical messengers within cells as well as agents of destruction. Indeed, since the 1970s, scientists have discovered that radical messengers help regulate blood pressure, stimulate immune cells to attack infectious microbes, and switch on self-destruct programs in deranged cells that might otherwise form tumors.

And here's a very important related issue: When levels of free radicals within cells rise—which is thought to happen, for instance, when we exercise hard—they act as signals to boost SOD and other natural defenses against free radical damage. Thus, raining antioxidants down on cells may make them drop their overall guard against oxidative stress, leaving them open to damage that more than offsets the possible gain.

It should be noted that this is a controversial idea. But a number of studies in recent years indicate that large antioxidant doses are risky, and that interfering with free radical metabolism can backfire. In fact, researchers who assembled what may be the most comprehensive look at the issue cited such interference as the most plausible explanation of their troubling results. Reporting in the February 28, 2007, issue of the *Journal of the American Medical Association*, the European team pooled results from forty-seven high-quality clinical trials with 180,938 participants and found that taking five popular supplements—vitamins E, C, and A, beta-carotene, and selenium—was associated with a statistically significant 5 percent increase in the overall risk of death. The risk, they found, was mainly associated with vitamins E and A and beta-carotene.

While hotly debated—and clearly not the final word on antioxidants—the *JAMA* report was consistent with other studies that have raised doubts about the wisdom of taking heavy doses of anti-oxidant supplements.* In 2005, four major clinical studies were published showing that vitamin E supplements not only failed to fend off killers such as cancer and heart disease but also appeared to slightly increase the risk of death from all causes. Mulling the spate of dismaying data, the *Annals of Internal Medicine* dubbed 2005 the "annus horribilis" for vitamin E.

All this has given skeptics about the prospects for life-span extension the best opportunity to say "I told you so" since the 1950s, when George Williams, the influential evolutionary biologist, seemed to permanently vanquish the search for anti-aging therapies to the junk-science heap. But it would be a mistake to dismiss companies like Eukarion as lost causes. As part of the first wave of scientifically credible companies to spring from research on aging, they were the Roald Amundsens and Robert Pearys of applied gerontology, firmly planting its flag in biotech's forbidding landscape. Eukarion, in particular, helped stir commercial excitement about aging science. And it's too early to write off Eukarion's medicines, which may yet reach the market for diseases of aging. While it's true that chronic, heavy-handed fiddling with free radical metabolism seems problematic in healthy people, judicious use of potent antioxidants like Eukarion's might ameliorate diseases involving severe oxidative stress.

The companies also pioneered a business model that has been

---

*While popping antioxidant pills has come under fire in the medical literature, there are still good arguments for taking multivitamins that contain modest amounts of antioxidant vitamins. Distinguished biochemist Bruce Ames, for example, argues that excessive consumption of nutrient-poor refined foods leaves many people in today's world with inadequate intake of a number of vitamins and minerals, possibly boosting the risks of cancer, neural decay, and accelerated aging.

emulated by similar start-ups that followed. The model's forte is to enable biotechs to benefit from buzz about aging science without getting lumped with the flaky, miracle-elixir crowd. The key is to grow orthodox drugs from the unorthodox seeds of anti-aging research. A true anti-aging drug should, almost by definition, avert or slow the progression of a wide array of degenerative diseases that arise late in life. That would give it colossal value, but not as a dubious youth preservative sold in health-food stores. Instead, its full value could be unlocked by developing it as a conventional prescription drug that could address a whole slew of major diseases—it would be akin to disease-preventing drugs that lower cholesterol and high blood pressure, only far more versatile. As we'll see, Sirtris Pharmaceuticals—the biotech start-up cofounded by resveratrol researcher David Sinclair—decanted this idea into entrepreneurial rocket fuel.

The nifty business model also finessed a problem that had long suppressed drug-industry interest in gerontology: the fact that neither government regulators nor doctors see aging as a condition warranting treatment. But industry players might eagerly pursue a compound whose potential uses include, say, warding off diabetes, arthritis, cancer, and various neurodegenerative diseases—not unlikely for a medicine that truly retards aging. Such a drug might well be prescribed to hundreds of millions of people across the globe. Over time, that would mean a large swathe of the population would wind up aging slower *as a side effect*. In short, Eukarion and its ilk managed to devise a way, inadvertently, to foment an anti-aging revolution without making any claims at all about life-span extension.

But what about this chapter's big question: Do free radicals really do us in?

The short answer is—well, unfortunately there is no short answer. The free radical theory is still very much alive, and in recent years it has undergone another major update by its current advocates—call it Denham Harman's Big Idea, Version 3.0. The update posits that as we lose our lifelong war with free radicals, oxidative-stressed-out cells turn into catatonic zombies that drool toxins—biologists call this spread-the-hurt state the senescent phenotype. This version also postulates that free radical damage abets chronic, low-level inflammation in our arteries, brains, muscles, and other tissues. Such insidious inflammation is thought to help prepare the way for the onset of major killers—cancer, heart attacks, diabetes, Alzheimer's disease—and then help them polish us off.

On the experimental front, however, the picture is murkier than ever. In 2005, University of Washington researchers reported that by genetically altering mice to produce extra catalase in their mitochondria—catalase, recall, is the antioxidant enzyme that eliminates hydrogen peroxide—they boosted the animals' life expectancy by 17 percent and their maximum life span by 21 percent. But other scientists, most prominently Arlan Richardson at the University of Texas, have compiled lots of data suggesting just the opposite—boosting rodents' native free radical defenses doesn't extend their lives, and in some cases shortens it.

Still, even doubters about the free radical theory acknowledge that the damaging molecules may be coconspirators in aging, and that mitigating their damage might help avert diseases of aging. If taking loads of antioxidants isn't the way to go, though, what is? That's still an open question, but several lines of research have suggested safe ways to do it. One is calorie restriction, or CR, which we'll take up later—as noted earlier, CR, which entails a drastic reduction in food intake, is the only intervention that has reliably retarded aging across many species. No source of insight on free radicals and aging, however, is more fascinating than the menagerie of weirdly long-lived animals that we'll explore next.

3

HAGRID'S BAT AND THE SABER-TOOTHED SAUSAGE

STUDDED BY MONUMENTAL rock formations towering over the Siberian forest, the Stolby nature reserve isn't the kind of place one associates with longevity. Stalin's Gulag sent many of its victims to a labor camp in the nearby city of Krasnoyarsk. Next to the reserve's entrance stands a memorial to an appalling number of people who have fallen to their deaths from its sheer granite pillars—Stolby is a magnet for Russia's most macho rock climbers, who get their adrenaline fixes by dispensing with ropes, protective gear, and sometimes even shoes. Yet it was here that the most remarkably long-lived animal known to science was discovered.

Russian wildlife biologist Alexander Khritankov made the find while trolling through a cave occupied by hibernating Brandt's bats, close relatives of the ubiquitous "little brown bat" we often see fluttering overhead on summer evenings. His predecessors at Stolby had banded more than fifteen hundred of the bats in the early 1960s. (Banding entails placing small metal rings embossed with identifying information around animals' legs.) Nicknamed Hagrid by U.S. colleagues because, like the Harry Potter character, he spends most of his time roving the forest, Khritankov has been embedded with Stolby's fauna for years. When funding was short during his earlier days at

Stolby, he even trapped sables in order to raise money for a computer and other essentials. Khritankov and a colleague had been taking a census of the banded bats every year or two; the animals hibernate from late September to June, giving ample opportunity to take stock. By 2005 there seemed to be only a single, rarely sighted member left. Coming across it one day, Khritankov discovered that it had carried its band for forty-one years.

The finding made it into the *Journal of Gerontology*, but it received scant attention in the media—who cared about a sleepy old bat? To gerontologist Steven Austad, however, Hagrid's bat seemed to have sipped from a pool of ambrosia deep in the taiga forest. It had lived ten times longer than similarly sized mammals, such as mice, thereby outrageously defying a longevity pattern that almost seems a law of nature. To wit: Among mammals, the larger a species in physical size, the longer its life span. You can see the pattern at a glance by considering the maximum life spans of captive mice, rabbits, lions, and elephants—roughly 4, 10, 30, and 60 years, respectively. The size-versus-life-span correlation has long fascinated gerontologists, and it seems to say something very basic about aging. But if it does, Hagrid's bat, which weighed less than half as much as a typical mouse, has something strange and important to add.

Austad, the University of Texas researcher we met earlier, is a specialist in comparative gerontology, the study of aging across species, and was a coauthor on the bat report along with one of his U.S. colleagues from Russia, Andrej Podlutsky. In 1991 Austad devised a measure, the longevity quotient, or LQ, that can neatly express just how far from normalcy Hagrid's bat is. LQ is defined as the ratio between the actual life span of a mammal and the life span it seemingly should have, based on the usual correlation between longevity and size. (For computing LQ, size is defined as a species' average weight, and life span is its greatest recorded longevity.) Thus, mammals that fit the usual pattern, such as rats, lions, and elephants, have LQs of about 1. Oddly short-lived species for their size have LQs less than 1—the white-eared opossum, which keels over by age four, has an LQ of 0.3. Mammals with anomalously long life spans have LQs over 1. Examples include porcupines (3.4), humans (4.5), and little brown bats (5.8).

At this writing, Hagrid's bat is the runaway LQ champ—it racked up a 9.8.

The bat's feat seems especially remarkable given that for so long he (it was a male) had surmounted the rigors of life in the wild. His hearing had remained razor sharp, otherwise he wouldn't have been able to detect insect prey with his radarlike echolocation system. He had maintained youthful zip and maneuverability in order to elude predators and catch prey. And as to mental alertness and strategic smarts, he had obviously stayed right on top of his game—there's very little room for error in the life of a Siberian bat. It was as if Roger Federer had dominated Wimbledon for more than a century.

Why do porcupines, people, and bats live much longer than they seemingly should? Gerontologists once thought cases of anomalous longevity basically represented accidents of nature. In 1957, evolutionary biologist George Williams proposed a far more interesting explanation. While laying out his influential theory on the evolution of aging, he argued that slow aging tends to evolve in animals that are very good at not getting eaten. Peruse the list of animals with high LQs, and you'll see that they all have especially strong defenses against predation. Few predators can catch a night-flying bat, for example, and porcupines keep predators at bay with quills.

Williams realized that evolution, which is all about maximizing reproductive fitness, ensures that predator-resistant animals are built to last. If they aged as fast as, say, mice, they'd miss out on a lot of baby-making chances afforded by their imperviousness to predation. In other words, when it comes to longevity, nature strictly adheres to the old saw about the rich getting richer and the poor getting poorer: Animals that are especially good at fending off death from predation (as well as other mortal risks such as cold, starvation, and infectious diseases) are blessed with bodies that age slowly, while ones at high risk of an early demise get stuck with bodies that quickly go downhill.

This reasoning handily explains why humans age much slower than other animals about our size. We brainy types can outsmart predators big enough to see us as food, and we devise tools and tricks to keep ourselves from starving or dying of exposure. It also illuminates why larger animals generally live longer than smaller ones—if you're relatively large, there are fewer predators that can pose a mortal threat to you. Get big and you also have more bodily reserves to stave off death when food is short or winter is on. So you're likely to have more offspring at older ages, which causes your combination of get-big and live-long genes to become more common in the next generation. Add an eon and a few eras to this Darwinian recipe and voilà, largeness has evolved along with durability.

But high-LQ animals haven't won the evolutionary jackpot. As highlighted by disposable soma theory, the evolution of longer life span entails a trade-off that lessens fertility—long-lived animals mature relatively slowly and produce smallish numbers of offspring. Of 133 species of bats whose fertility was examined in one study, nearly two-thirds were found to have only one pup at a time—mice can have more than ten pups per litter. In fact, the best predictor of long life span among bat species is a low reproductive rate.

Such observations gave Williams's explanation the ring of truth. Still, it had the slightly squishy feel of a theory whose predictions haven't been experimentally tested—and might well never be. What rational scientist would dream of trying to observe evolution in the act of shaping a mammalian species' LQ? (As explained below, measuring LQ makes sense only for warm-blooded animals.) Even for a short-lived rodent, you'd probably need to track many generations over centuries to detect a significant change in life span, assuming that such change was in the cards. Forget it—the very idea of registering the evolution of mammals' life span is laughable.

But Steve Austad wasn't so sure it was out of the question. And if any scientist had the right stuff to do something laughably improbable, it was he. Before going into gerontology, he'd spent time as a rambling pool shark, appeared in drag on prime-time TV, and come close to death in an accident involving a lion and a suicidal duck.

. . .

Austad was tapping away on his computer when I arrived at his top-floor office at the Barshop Institute for Longevity and Aging Studies, a cluster of gleaming new buildings atop a wooded hill ten miles west of San Antonio. La Quinta Inns founder Sam Barshop donated $4 million to launch construction of the $20 million center, an offshoot of the University of Texas Health Science Center, and some of the biggest names in gerontology have joined its faculty since it opened in 2005. (Austad was recruited from the University of Idaho soon after it was formed.) Dressed in jeans and sneakers, he has the solid build, and the reassuring but watchful demeanor, of a man who used to train lions for a living—which he did. I spotted a hawk floating over the rolling olive hills out his window, and when I commented on the spectacular view, he mentioned that cougars had occasionally been sighted below at dusk, regally strolling onto the institute's manicured grounds before shying back into the brush.

Big cats, with their ho-hum LQs, don't fascinate him the way they used to. Today he's enchanted by bats and other peculiarly long-lived creatures, such as bizarre little rodents called naked mole-rats. But before delving into his research on the high-LQ elite, I couldn't resist asking about stories I'd heard of his former life as a lion trainer. Did he run away from home to join the circus or what?

Actually, it turned out it was his dad who ran away from home. Seized by wanderlust when Austad was ten, his father, a newspaper pressman in Los Angeles, bought a trailer, loaded up the family, and set out on a zigzagging five-year tour of the country. "He belonged to a strong union and had a traveling card," Austad told me, "so it was easy for him to get temporary jobs at newspapers. We'd spend two months here, and three months there." Eventually they circled back to L.A., where Austad finished high school, entered the University of California at Los Angeles, and perfected his eight-ball game.

The Vietnam War was raging when he graduated in 1969 with a BA in English literature, and Uncle Sam wanted him. "I was quite the sixties idealist, so I told my draft board to stick it," he said, grinning, clearly unrepentant and still very much the son of his unconventional father. "Then I basically went on the run and lived an underground existence for several years. I made money hustling pool. I was a taxi driver in New York for a while. Ironically, I even worked for a few weeks as a clerk typist on a U.S. Army base while hitchhiking around Europe."

The FBI eventually located him in Portland, Oregon. But before he could be arrested, he discovered that his draft board, apparently apoplectic after being told to stick it, had failed to follow required procedural steps when trying to nail him, fatally undermining the government's ability to prosecute him as a draft dodger. Returning aboveground, he joined forces with his karate instructor, who owned and trained lions. Austad was well equipped to fool with the king of beasts—he'd always loved animals, was unflappable, and had a youthful inability to perceive his own mortality. He soon found himself working with lions in Hollywood. "That's when I got very interested in how animals tick," he said. "Because lions are so social, they're easier to train than other big cats. They give you a lot of facial and postural cues about what they're thinking. Of course, they're always testing you to see who's dominant, but you can tell when they're getting really upset and get out of the way. Bears and tigers are much harder to read."

He didn't always manage to get out of the way. For an episode of the 1970s TV series *The Bionic Woman* he was hired to pose as the female heroine under attack by a lion. Apparently excited by his odd behavior and getup—a skirt and wig—the cat got out of hand during the scene and "knocked me half goofy," he recalled. "Then it tried to mate with me. By the time they got him off me, everybody on the set was roaring."

Only his pride was hurt. But he wasn't so lucky the next time one of his charges took him down. "I was walking a lion on a chain in this large, enclosed area where we kept them," he said, "and suddenly a duck ran out from a clump of rushes right in front of us. The lion jumped on the duck before I could react. Lions get very possessive about their food, so I should have just backed off at that point. But in that split-second I saw it as a dominance issue, and so I told the lion to drop the duck and whacked him over the head with the chain. He came after me and got me down." Alert coworkers quickly intervened, but not before the cat had chomped Austad's knee.

The episode didn't end his lion-training days, but it helped convince him to rethink his career. Not long after, he entered graduate school at Purdue University, planning to study lions in the wild. But fiercer creatures beckoned—he wound up doing his Ph.D. research on spider combat. Then he switched to studying sex, but while investigating the mating behavior of wrens in Venezuela, a chance discovery inspired him to launch the first, and so far the only, major field study on the evolution of mammalian life span. The serendipitous finding occurred when a colleague who was trying to put radio collars on tropical foxes repeatedly found his traps filled with opossums. Making the most of the situation, he and Austad began monitoring the inadvertently captured animals and soon discovered that they age at lightning speed, often turning from vibrant adults to burned-out cases over mere months—no one had known opossums get old so fast. Pondering this oddity, Austad came up with a clever idea for sneaking a peek at the blind watchmaker tinkering with longevity.

"Ecologists have known forever that predators are often scarce on islands," he explained. "I assumed that that had repercussions on life span." That is, if Williams's hypothesis about the evolution of life span was right, prey animals on predator-free islands should evolve abnormally long life spans, for they have the best protection of all against predation. He also reckoned that this effect could be readily seen in opossums. Not only do they normally age rapidly, but also, unlike most short-lived prey animals, they're big enough to carry radio collars for extended monitoring. Austad soon found the place to go: Sapelo Island, a sparsely populated enclave five miles off the coast of Georgia that's crawling with ticks, rattlesnakes, and opossums—but is free of foxes, bobcats, high-speed roadways, and other opossum killers. Further, it was known that Sapelo had been separated from the mainland for only about four thousand years, which meant he would be studying fresh evidence of life-span evolution, if any was to be found.

"When I got to the island," he recalled, "the first opossum I saw made me think, 'These are the animals I want.'" Unlike furtive, nocturnal opossums on the mainland, "it was walking across the road in the middle of the day," as if utterly oblivious to predators. "I jumped out of my truck, grabbed it, and slapped on a radio collar. I was able to walk right up to others while they were sleeping on top of the ground."

After three years of tracking opossums on both Sapelo and the nearby mainland, his data showed that the island animals' average and maximum life spans were dramatically longer—23 percent and 45 percent, respectively—than those of their relatives across the water. And in keeping with the disposable-soma trade-off that the Sapelo opossums had made, their average litter size was 26 percent smaller. Darwin, whose own island studies figured prominently in *On the Origin of Species*, would have loved it: Austad had managed to register life-span elongation on Sapelo Island just as predicted by the evolutionary theory of aging.

• • •

One of the little surprises of the anti-aging quest is the fact that miracle-elixir peddlers have never brought out products containing essence of bat, porcupine extracts, or island-opossum hormones. The yuck factor probably isn't the reason. After all, rejuvenation hucksters have convinced their desperately credulous customers to buy into everything from monkey-testicle implants to gulping down their own urine. Apparently the hucksters haven't been reading the longevity-quotient literature. If they had, we'd doubtless be bombarded with spam about sleek, beefy ninety-year-olds mixing freeze-dried bat powder into their carrot juice.

Serious anti-aging researchers have good reasons to study high-LQ animals, though. If humans' LQ were as high as that of Hagrid's bat, we wouldn't get elderly until we were well over 200, and some of us would make it past 250. How do such animals do it? Unfortunately, we know surprisingly little about their special genes or other peculiarities at the molecular level. That's because gerontologists have fixated on the same short-lived "model" animals that other life scientists have long favored—mainly nematodes, fruit flies, and mice.

In recent years, Austad and other comparative gerontologists have begun to correct this knowledge deficit. They haven't yet pinpointed the sources of ultra-high-LQ magic. But they have shaken up the field with a series of provocative findings.

One of Austad's first major studies helped trash a hypothesis that had dominated thinking about animal life spans since the nineteenth century. Known as the rate-of-living theory, it basically holds that if you live fast, metabolically speaking, you die young. The idea had a lot going for it. Scientists had long known that metabolically sluggish, cold-blooded animals such as turtles and lizards tend to live longer than similarly-sized mammals. This made a lot of sense to ninteteenth-century thinkers, to whom mammals seemed like factory machines that tended to break down relatively fast because they ran hot at high speed. The theory also handily explained why insects and other cold-blooded animals live longer when maintained at chilly temperatures—the cooling slows their metabolic rates, which in turn brakes their aging. (This is the main reason that calculating LQ for cold-blooded animals doesn't make sense—it often would say more about the temperature of their habitats than their aging process.)

German physiologist Max Rubner fleshed out the rate-of-living theory during the early 1900s with experiments suggesting that all animals use about the same amount of energy per gram of body weight during their lives. To get your mind around that, consider this: Because all animal cells are similarly constructed, it makes sense that their molecular machinery would burn out after exposure to similar amounts of inner fire, regardless of the type of body that the cells are in—that's essentially what Rubner seemed to have discovered. His finding fit nicely with the fact that relatively big, slow-aging mammals have slower metabolisms. If the cells of a rhinoceros burned energy at the same rate as those of a mouse, the rhino would need to maintain a surface temperature of more than 212°F, the boiling point of water, to dissipate the heat generated within its body. No wonder a rhino's metabolism runs comparatively slowly, spreading its cells' allotted lifetime calorie expenditure over four to five decades. The hyperkinetic mouse, which has a lot on its plate and a tight deadline, runs through approximately the same energy allotment per cell in less than four years.

The rate-of-living theory seemed to explain just about everything about aging. American biologist Raymond Pearl, a brilliant and famously doctrinaire scientist, maintained that it even accounted for the fact that people whose jobs require heavy physical labor, such as coal miners, tend to die earlier than those whose professions don't force them to burn energy as fast, such as lawyers. (Raise your hand if you'd like to see Pearl's hypothesis rigorously tested by placing a control group of lawyers in a coal mine.) The headline on a 1927 article Pearl wrote for the *Baltimore Sun* made this point in a way guaranteed to thrill adolescents and goldbricks everywhere: WHY LAZY PEOPLE LIVE THE LONGEST. (Pearl, by the way, died in 1940 at sixty-one, his life possibly cut short by his heavy expenditure of energy on browbeating colleagues.) While Pearl's rate-of-living idea about the perils of hard work was highly dubious, to say the least, later studies with insects lent some limited support to it. My favorite, although it had a sad ending, was the one in which honeybees were forced to carry around little weights that made them work harder than ever. They died young.

Rate-of-living theorists may have felt all-wise on aging, but they had to resort to hand waving on the issue of why expending energy withers our tissues as we age. Unlike factory machines, most cells are self-renewing. Thus, it was unclear why ones that were somehow damaged by their energy-using process wouldn't simply be replaced by vigorous new cells.

The theory's proponents also had a bat problem: The longevity of flying mice, as Germans call bats (die Fledermaus), just didn't compute. Although life-span and metabolic data on bats were spotty before the 1980s, scientists had a pretty good idea that their cells burn much more energy over their lives than other species' cells do—a glaring violation of the rate-of-living theory. Some other peculiarly long-lived species, including various kinds of birds, also seemed capable of sustaining high metabolic rates far longer than allowed by the theory. In 1991 Austad and a colleague highlighted the problem with a sweeping study of mammalian life spans and metabolic rates, including a compilation of the LQs of 580 species, of which 50 were bats. Among other things, it showed that during their lives, bats burn more than four times as much energy per gram of body weight, on average, as do marsupials, a low-LQ group that includes opossums. When Austad got finished with it, the James Dean theory of aging—live fast, die young—was looking quite elderly and in very delicate health.

Still, Harman's related free radical theory, which also tied aging to energy expenditure, remained vibrant. Unlike the simplistic rate-of-living idea, it could explain why expending energy leads to aging, as well as why small animals with fast metabolic rates don't live very long: Their cells generate free radicals at a higher rate than do those of larger animals, hence more quickly accumulate irreparable free radical damage to DNA and other key molecules. In principle, the free radical theory could also explain the strangely slow aging of bats, birds, and other high-LQ species. Presumably, they have extrastrong free radical defenses, enabling their bodies to burn through more calories before accumulating fatal damage. It followed that in order to pinpoint the anti-aging mechanisms responsible for such animals' longevity, scientists should focus on how they deal with free radicals. Not surprisingly, that goal has dominated research on high-LQ animals for two decades.

Examining such species' natural antioxidants was an obvious first step. Surely long-lived animals have high levels of the free radical blockers in their tissues. Some early data on the issue agreed with this appealingly simple concept. In 1980, researchers reported that the activity of superoxide dismutase, the potent antioxidant enzyme that inspired Eukarion's drugs, rises in tandem with various species' lifetime energy expenditure. But a more thorough 1990 study led by Rajindar Sohal, then at Southern Methodist University, failed to show a correlation between overall levels of key antioxidants and life span in six mammalian species—mice, rats, guinea pigs, rabbits, pigs, and cows. Similar research by Gustavo Barja and colleagues at the Complutense University of Madrid in Spain led to an even bigger surprise: The longer a species' life span, the lower its antioxidant levels. That was just the opposite of what you might expect, especially if you were sold on the idea of taking antioxidant supplements.

Puzzling over these data, free radical theorists saw a way to explain them without tossing out Harman's hypothesis: It must be that long-lived animals spew fewer radicals than short-lived ones do when their cells burn fuel, hence they can afford to maintain relatively low levels of antioxidants. The idea stood to reason. If an animal is geared to live a long time, it would likely possess a highly efficient way to minimize free radical damage, and simply cutting production of radicals would probably be more cost-effective than maintaining a souped-up antioxidant system. Checking it out, Sohal and Barja independently compared free radical production in two like-sized, warm-blooded animals with comparable metabolic rates: rats, which expire before age five, and pigeons, which have maximum life spans of thirty-five years (think of that the next time you call them flying rats). The expected pattern was duly demonstrated: Both researchers reported evidence that pigeons have cleaner-burning mitochondria than rats.

Bats appear to have evolved the same trick. In one study, heart mitochondria from little brown bats, which have been reported to live to thirty-four, were found to be at least twice as efficient as those of short-tailed shrews, whose maximum life spans are two years, in terms of free radical production per unit of oxygen consumed.

No one thinks that efficient mitochondria alone confer high LQ. Insect-eating Mexican free-tailed bats, for example, get a whopping 60 percent of their calories from fat, yet somehow their arteries don't get clogged—their longevity probably involves special lipid handling. Fruit-eating bats can ingest huge amounts of sugar without becoming diabetic, and they also function perfectly well with extremely low blood-sugar levels. Thus, the evolution of long life apparently involves at least somewhat different sets of defenses tailored to the hazards of different niches. Still, the slow-aging tricks reported most often in high-LQ animals appear to mitigate free radical damage, and for many years the animals' main message to gerontologists seemed to be, "It's the free radicals, stupid." But that was before the naked mole-rat had its surprising say.

• • •

The term *naked mole-rat* apparently came from the same language factory that gave us *driveways* for parking and *parkways* for driving. The mouse-sized mammals are neither mole nor rat, belonging instead to the rodent subgroup that includes guinea pigs and porcupines. The word *naked* is also a bit misleading—while lacking fur, they have sensory hairs all over their pinkish-gray bodies. The dual-beast designation, however, does convey their vague similarity to the chimera of Greek mythology, a monstrous conjoining of a lion, goat, and snake.

After seeing naked mole-rats at zoos, people come away exclaiming to friends that they saw these weird things that looked like a colony of saber-toothed sausages, or maybe mice implanted with walrus genes. Both are perfectly accurate depictions.

NMRs' first claim to famous strangeness was the discovery that they are an extremely rare mammalian version of the social insects, such as honeybees and termites, which live in colonies populated by workers that act as a support system for a single breeding queen. In recent years, another NMR peculiarity has come to the fore—their extreme longevity. The leading authority on that topic, Rochelle Buffenstein, works just downstairs from Austad at the Barshop Institute outside San Antonio.

When I visited Buffenstein one afternoon, she took me to the building's basement to see some of her mole-rats. Walking into a dimly lit room, I found myself surrounded by scores of chirping NMRs merrily tending their nests inside clusters of shoebox-sized containers connected by clear plastic tubes. To my surprise there was no airtight barrier sealing them off from the outer world's germs. NMRs are so hardy, Buffenstein explained, that she's never felt the need to keep her animals inside the kind of sterile bubble commonly used for research rodents. As if to highlight that point, she plucked one out of its colony and handed it to me.

NMRs are extravagantly ugly, but in person they have Jimmy "the Schnozzola" Durante's semigrotesque charm. The one I held was remarkably docile, squirming a little, but unlike the mice I've known, it showed no propensity to bite. That was nice, as I'd heard that the strong-jawed, tuber-chomping rodents could bore through concrete with their teeth. It had the bald, wrinkled, buck-toothed, querulous, squinty-eyed look of a slightly demented codger born well before the age of orthodontia. In fact, it was truly ancient. Its exact age wasn't known—it had been caught in the wild—but Buffenstein estimated that the animal, the "old man" of her colonies, was pushing twenty-nine. It suddenly dawned on me that this was probably the oldest rodent on the planet, and possibly the longest-lived one ever. I gingerly handed it back to her feeling as if I had been momentarily entrusted with a two-thousand-year-old Han Dynasty vase.

Mole-rats were introduced to science by German naturalist Eduard Rüppell in 1842. The son of a German banker, Rüppell explored NMRs' northeast Africa habitat during the ninteeenth century's golden age of heroically traipsing the biosphere and bringing pieces of it home—he bagged everything from a vulture with a wingspan of nearly nine feet to a diminutive desert fox. Despite NMRs' enticing formal name, *Heterocephalus glaber* (meaning, to a first approximation, weird-headed baldy), they remained little-studied curiosities until the 1960s, when Jennifer Jarvis, a Kenya-reared daughter of English missionaries, selected them for her Ph.D. research in zoology at the University of East Africa in Nairobi. She began by studying the anatomy of their unusual ears, but soon became interested in other oddities, such as the fact that only one female in each of her captive NMR colonies appeared to be making babies—a pattern that made no sense at all for a rodent. Were they too stressed to reproduce, or what?

Meanwhile, a remarkable coincidence was shaping up: In the mid-1970s, University of Michigan entomologist Richard Alexander theorized that mammals might exist that form insectlike social colonies, each with a single reproducing queen. One day a colleague who heard about his hypothetical beasts mentioned that NMRs seemed to fit the bill. "What the hell is a naked mole-rat?" Alexander reportedly asked. Three years later, he joined Jarvis and another NMR fancier, Paul Sherman of Cornell University, for a trip to Kenya to capture NMRs for laboratory colonies. Jarvis, who had suspected the rodents might be termitelike, went public with the idea in 1981 after talking with Alexander and Sherman.

Since then NMRs have become international celebrities—there's even a mole-rat Disney character named Rufus. The fascination is warranted. For one thing, they use their teeth to dig burrow systems that can be nearly two miles long. Colonies of up to three hundred NMRs organize themselves around large-bodied, domineering queens that mate with one to three consorts and typically produce many hundreds of babies during their lives. A queen keeps workers in line by continually—and literally—pushing them around. When she encounters one of her subjects in a tunnel, she shoves it backward or walks over it, and her aggression appears to help suppress the groundlings' fertility and make them work hard. Rarely shoved workers in tunnels that are remote from the queen's chamber tend to goof off. Sometimes they even become reproductively active—a huge no-no from the queen's perspective that can lead to deadly showdowns.

Younger NMRs do most of the pup care and tunnel maintenance. Older ones take on hazardous jobs, such as kicking dirt from newly dug tunnels out surface openings—birds, snakes, and other predators often pick off these "volcanoers" as they pile up molehills. When a volcanoer encounters a snake, "it rushes off toward the nest 'screaming,'" according to one report. (NMRs have eighteen known kinds of vocalizations, including a latrine-assembly call, aggressive chirps and trills, and hisses that mean "Let's kill the intruder now.") A worker's screams summon defenders, who rush toward the snake hissing and snapping their formidable teeth—their preferred defense is to quickly wall off intruders, but they've been known to rip a snake apart before it realizes that you don't mess with saber-toothed sausages.

Buffenstein learned about NMRs as a student at the University of Cape Town in South Africa, where she worked as Jarvis's research assistant. A native of Zimbabwe, she'd grown up on a farm there, dreaming, as a youngster, of becoming a veterinarian. But in college she got hooked on research, largely as the result of a 1980 field trip to Kenya with Jarvis to capture mole-rats. They brought back an entire NMR colony of nearly three hundred animals, grabbing them by hand one at a time in a sweet potato field. "It was hard work," Buffenstein recalled. "We'd scrape the soil down [to make a breach in their tunnels] and then sit there with a hoe. When a worker came to investigate the breach, we'd bring the hoe down behind it," preventing its retreat so as to grab it. "Once I stuck my hand in a tunnel and came across a dead snake. I nearly died. There's nothing more frightening than putting your hand in the ground and feeling something cold and slimy."

When Buffenstein launched her research career in physiology a few years later, Jarvis gave her some NMRs to study from the colony they'd captured in Kenya. Trilling away in plastic containers in her office, the animals gradually took over her professional life. Over the following decade she detailed many aspects of their basic metabolism, including one of their most unusual features: NMRs have largely lost the ability to regulate their body temperature—they're basically cold-blooded mammals. This is probably a result of their living for millions of years in subterranean nests that remain at a nearly constant temperature year-round.

As the years passed, her NMRs' longevity became the oddity that stuck out most. By mid-2001, her oldest animal was making history every day—approaching his twenty-eighth birthday, he had outlived the previous holder of the rodent longevity record, an Asian porcupine that had lived for twenty-seven years and four months. While it's generally considered bad form in science to give pet names to lab animals, exceptions are made for ones that set longevity records. As word got around about the extraordinary animal, one gerontologist proposed dubbing him "Milton the Mole-Rat," a nice, old-fashioned name that was easy to remember. But, like all of Buffenstein's animals, the oldest NMR had long been identified by a number—it happened to be 007, and so among those who knew him well he was known, of course, as James Bond.

While 007 wasn't quite as suave as his namesake, he did have the makings of a sexy bon vivant. As the senior consort of his queen, he continued siring pups right up to his death in April 2002 at age 28.3. It wasn't clear what finally brought him down. A postmortem showed no signs of cancer or other diseases that typically kill rodents in life-span studies. He did look old and tired—his skin resembled thin, translucent parchment. "I think he just gave up and stopped eating," Buffenstein said. At death his longevity quotient was an estimated 5.3, not large compared with the 9.8 achieved by Hagrid's bat, but monumental for a rodent.

Galvanized by 007's outlandish durability, Buffenstein, then at the City College of New York, pushed full-bore into studies on NMR aging. The results soon included some of the decade's hottest discoveries in gerontology. The first one, published in 2002, suggested that NMRs basically don't age during their first two decades. Examining the animals at 5, 10, and 20, her team found no age-related changes in bone mineral density, body mass, body-fat content, or basic metabolic rate, which is determined by measuring oxygen consumption at rest.

Later studies yielded more evidence that the animals Buffenstein had helped pull out of Kenya's sun-baked soil more than two decades earlier hadn't really aged in the normal sense of the word. She found that the queens remain fertile throughout their lives—one bore and suckled more than nine hundred babies during her twenty-four-year life. NMRs' blood vessels stay youthfully elastic as they get older, helping to fend off heart disease. The animals appear to be utterly immune to cancer, which often kills mice. Most stunning of all, their death rate doesn't rise rapidly with age as it does in other mammals. In fact, Buffenstein discovered that her NMRs are no more likely to die toward the end of their lives than they are as strapping youths.

They do finally die, of course. But like 007, says Buffenstein, NMRs often expire without having shown any sign that they are sick or dying. Putting these extraordinary findings together, in 2008 she proposed that NMRs are the first mammals that show "negligible senescence." (Gerontologist Caleb Finch, you may recall, earlier highlighted this phenomenon, and coined the term for it, in fish and some other animals that show little or no bodily deterioration with the passing of time.) NMRs may not be immortal, but you wouldn't know it from looking at them—at least for about thirty years.

Mole-rats' longevity may be shocking, but in a way it isn't surprising. It simply represents another example of the rich-get-richer-and-the-poor-get-eaten rule. Like bats and turtles, NMRs have a strong defense against predators—living underground. Their African habitat's nearly constant temperature eliminates another major killer of small rodents, winter cold. And Buffenstein believes that living in highly cooperative societies has further lowered their mortality risk by enabling communal care of young, team foraging, food sharing, and the "intergenerational transfer of information"—young mole-rats probably learn risk-reducing tricks, such as making a certain sound when an intruder enters the nest, from their elders. Similarly, evolution has bestowed extralong life on social-insect queens, some of which can live nearly thirty years (nope, that's not a misprint) within the low-mortality zones of underground nests, surrounded by myrmidons ministering to their every need. It's likely that humans also evolved a high LQ partly due to social behaviors that lower mortality risks, including intergenerational transfers.

Explaining the evolutionary reason why NMRs live so long, however, doesn't shed light on how they do it. The mystery deepened when Buffenstein's group calculated that during their lives, the animals expend more than four times as much energy, ounce for ounce, as mice do. Even high-LQ humans don't come close to mole-rats' lifetime expenditure of energy per unit of weight. Given this, you might think mole-rats had evolved superpotent mechanisms to protect against free radical damage.

But Buffenstein found little evidence of that. For one thing, NMRs have unusually low antioxidant levels. In fact, the researchers discovered that one key antioxidant, glutathione peroxidase, is seventy times less active in the livers of mole-rats than in those of mice. That was only moderately surprising—recall that high-LQ animals like bats and birds are thought to have clean-burning mitochondria that release relatively few radicals, obviating the need for high levels of antioxidants. More surprising, though, when Buffenstein and colleagues closely examined the functioning of NMRs' heart mitochondria, the cellular power plants didn't seem to be especially clean-burning. Other data showed that NMR tissues are, in fact, riddled with free radical damage. Their lipids, proteins, and DNA—the basic constituents of cells—were found to exhibit two to eight times more free radical damage than the same molecules in mice.

Once again the African rodents had proved freakier than anyone expected. They were like badly rusted winter beaters somehow continuing to chug along year after year. And they were dragging along the battered free radical theory of aging like a falling-off muffler.

• • •

The discovery that slow-aging NMRs carry heavy loads of oxidative damage may have troubled free radical theorists, but it promised new leads in the hunt for anti-aging drugs. Elucidating this strange phenomenon hasn't been easy, though. NMRs are such lavishly quirky creatures that unraveling their high-LQ secret is like trying to grasp an enigma wrapped in a riddle hidden in a fanged hot dog. Still, NMRs have the same basic set of genes and physiological processes that other mammals do, so it's likely that at least some of their durability enhancements exist in other anomalously long-lived mammals, such as bats. Actually, it's more than likely—Buffenstein, combining forces with Austad and other researchers, has already found some highly intriguing commonalities between NMRs and long-lived bats.

First, both have strikingly low metabolic rates when they're not active. NMRs operate at relatively low body temperatures, about 91°F, versus 96°F or so among other rodents, and their resting metabolic rate is only about two-thirds that of mice—not surprising for underground mammals that are nearly cold-blooded. In keeping with their low metabolisms, NMRs have remarkably low fasting blood sugar and insulin levels. Similarly, Egyptian fruit bats' fasting blood glucose is less than half that of humans. And vampire bats' blood sugar runs so low that if the animals fast for more than a couple of days, they drop dead. No wonder Dracula always looks wan.

In light of these data, it probably isn't a coincidence that calorie-restricted mice and other rodents, which live up to 40 percent longer than usual, have lower body temperatures and blood insulin levels than rodents on standard diets. And among men participating in the Baltimore Longitudinal Study of Aging, the longest-running study of aging in the United States, those with lower body temperatures and lower blood insulin levels have tended to live longer. These two factors are among the handful of "biomarkers" that best predict a healthy, long life.

All this might lead you to think that the rate-of-living theory is correct after all—that being geared for low energy expenditure really is the secret of slow aging. But again, that's too simple. NMRs' metabolic rates can shoot up by 500 percent as they frantically dig tunnels right after a rain, rapidly fanning out through temporarily moisture-softened soil in search of the scarce tubers that sustain their colonies. Bats can rev even higher, increasing their metabolic rates by some 2,000 percent during flight. And recall that these high-LQ species expend far more energy per unit of body weight over their long lives than shorter-lived animals do. So the lazy life isn't necessarily a long one. In fact, there's reason to think just the opposite is true for our species, among others: The combination of low resting metabolism with high revving capability in NMRs and bats resembles the slow resting pulse and large pulmonary capacity of people who spend a lot of time vigorously exercising.

Another intriguing parallel between mole-rats and long-lived bats came to light in a pair of studies that appeared in 2006: The cells of both are strongly resistant to the toxic effects of chemicals that induce free radical damage. In one study coauthored by Buffenstein, blood-vessel cells from NMRs were utterly unfazed by doses of hydrogen peroxide that were triple the doses that induced signs of a mass die-off of similar mouse cells in culture. In the other study, overseen by the University of Michigan's Richard A. Miller, skin cells from little brown bats proved to be about four times more resistant to the lethal effects of peroxide than those of mice.

Clues about the roots of this cellular hardening emerged in 2006, when Buffenstein and an Australian colleague, Anthony J. Hulbert at the University of Wollongong, analyzed lipids that make up cell membranes in mole-rats and mice. They found that membranes from muscle and liver cells of mice had nine times more DHA, a fatty acid that's highly susceptible to free radical damage, than did comparable NMR membranes. Such a huge difference in fundamental body composition between two mammals is rare, and about a decade ago Hulbert devised a theory to explain why it can occur. Roughly, his "membrane pacemaker" hypothesis holds that membranes laced with DHA are literally more fluid, which abets a high level of metabolic activity in cells. Our busy neurons, for instance, are rich in DHA, as are hummingbirds' flight muscles and rattlesnakes' tail-shaker muscles. The downside of DHA-rich membranes is that they are easily oxidized. Thus, as in mole-rats, membranes in long-lived animals tend to have relatively low amounts of DHA compared with those of shorter-lived animals, according to Hulbert.

Austad and Buffenstein have discovered that proteins in mole-rats and bats are remarkably tough too. Asish Chaudhuri, one of their colleagues at the Barshop Institute, paved the way for the finding by inventing a novel way to measure protein "unfolding." The term refers to the fact that protein molecules are basically strings of beadlike amino acids folded into origamilike shapes that are critical to their proper functioning. Stresses such as heat and oxidizing chemicals can make them unfold. Then they may refold into dysfunctional shapes or, if the stresses are severe, totally fall apart—this "denaturing" process occurs, for instance, when raw eggs are heated in a frying pan. Chaudhuri's assay has revealed that proteins from NMRs' liver cells are about ten times more resistant to unfolding than such proteins from mice—a dramatic difference whose magnitude suggests it is an important contributor to mole-rats' overall durability. Proteins from long-lived bats are also remarkably resistant to unfolding.

The Barshop scientists have found evidence that unusual "chaperone" proteins are behind the remarkable toughness. Chaperones' job is to prevent other proteins under stress from misfolding or falling apart. The researchers discovered that when blood proteins from cows are bathed in extracts from bat or mole-rat cells, the bovine proteins become more resistant to unfolding—an indication that the high-LQ animals' chaperones in the extracts are lending a hand to the cow proteins. "My guess is that [NMRs and bats] have amazingly effective chaperones," said Austad. "It's especially interesting that these two very different species seem to have evolved a very similar trick. This is the single most exciting thing I've found since I've started doing comparative gerontology."

• • •

Flying home from the Barshop Institute, I began thinking that someday I may regret poking fun at the idea of freeze-dried bat powder. As if to show that my day of regret isn't far off, in early 2008 researchers reported implanting a gene that regulates limb formation from fruit bats into mice, engendering mice with significantly elongated forelimbs. If Austad and Buffenstein pinpoint a chaperone from bats or NMRs that keeps proteins pristine, it probably wouldn't take scientists long to stitch its gene into the mouse genome to see whether it produces mice with elongated life spans. If it did, the next step on the way to bat powder would be obvious—bioengineers have been making protein drugs from particular genes for thirty years.

But before our imaginations run away to the patent office, let's take a cautionary glance at the emerging big picture on high-LQ animals. The first thing worth noting is that despite some exciting similarities, different long-lived species have different bags of tricks to boost life span—not surprising, given that they evolved with vastly different life histories in widely disparate niches. Bats and birds, for instance, appear to slow the accumulation of free radical damage with high-efficiency mitochondria. Mole-rats, on the other hand, are remarkably rusted from their first chirpy days, yet remain resolutely chipper until suddenly dropping dead around age thirty from mysterious causes. Thus, it's unlikely that studying high-LQ animals will yield a single magic bullet to retard aging. And some anti-aging tricks may have limited effectiveness outside the niches and internal metabolic milieus that gave rise to them.

Still, metabolic idiosyncrasies of superhigh-LQ species clearly overlap with factors that we know are linked to human health and longevity, hence arguably point to especially promising directions in the anti-aging quest. The low blood sugar and insulin levels observed in long-lived bats and NMRs—as well as in calorically restricted animals, and in healthy humans during their later years—suggest that anti-aging researchers should scan the horizon for drugs that keep sugar metabolism under tight control. As we'll see later, such medicines are already shaping up as promising therapies to slow aging.

Perhaps the most important take-home message from research on high-LQ animals is how surprising they are. Who would have expected a buck-toothed burrower to age like a demigod, or a tiny bat to live longer than the average human did before the twentieth century? When it comes to aging, as with all else in the living world, evolution has generated "endless forms most beautiful and most wonderful," as Darwin so nicely put it. Hold that thought while reading the next two chapters—they tell the story of gerontology's biggest surprise of all.

4

THE GENES THAT COULDN'T BE

WHEN PLAYING ORACLE, great minds often contribute more to the human comedy than to the sum of knowledge. Nineteenth-century physicist William Thomson, better known as Lord Kelvin, is revered for his seminal work on thermodynamics and electricity; we also have him to thank for his wonderfully boneheaded pronouncement in 1895 that heavier-than-air flying machines are impossible—eight years before the Wright brothers took off at Kitty Hawk. Not to be outdone, Thomas Edison opined in 1922 that the "radio craze will die out in time." And in 1934 Einstein said that splitting the atom was scarcely doable—like "shooting birds in the dark in a place where there are only a few birds." German scientists split the atom in 1938.

Authorities on aging have added to the comedy, of course. One seminal contribution was George Williams's declaration in 1957 that the anti-aging quest is doomed to failure. Before we get to the momentous discovery that undercut his argument, let's briefly go over the main reasons he and many other scientists were so confidently pessimistic.

When Williams asserted that anti-aging research is like the hunt for perpetual motion, he rested his case on evolutionary theory. But what really gave his words weight was the fact that, just as his theory predicts, many things often do go wrong as we get old, as any doctor treating elderly patients knows. Highlighting this point, S. Jay Olshansky, who studies aging at the University of Illinois at Chicago, has calculated that if we were able to totally eliminate heart disease, U.S. life expectancy would rise by only about three years. A miracle cure for all cancers would yield roughly the same gain. The reason that the gains would be so small is that the risk of many killer diseases soars after age sixty. Thus, even if we all avoided, say, heart disease, other things would soon get us. (Turning this point around yields one of this book's main take-home messages: The only practical way to achieve a substantial life expectancy gain at this point is to develop anti-aging drugs that lower the risk of all diseases of aging at one fell swoop. That is, boosting longevity today mainly requires adding years in later life—the low-hanging fruit obtained by cutting childhood mortality has largely been plucked. But as I'll argue later in more detail, our current strategy of developing palliatives for degenerative diseases of aging, and applying them one at a time after much of the damage the diseases cause is already done, has become an expensive game of diminishing returns. We need to change the rules of the game, which is now effectively dominated by Williams's saturnine insight about the aging process. Anti-aging therapies would do that.)

Early studies on the genetics of aging supported the view that our bodily decline is intractably complex. In fact, they suggested that hundreds to thousands of genes team up to push us off a cliff. One estimate came from the lab of Michael Rose, the evolutionary biologist at the University of California at Irvine who generated long-lived fruit flies by selective mating. When he and colleagues analyzed the genetic changes behind their flies' longevity, they concluded that about 2 percent of the insects' genes—perhaps two hundred to four hundred— were involved in the control of aging. Since our genome is larger and more complex than that of flies, it followed that a somewhat larger number of genes, perhaps five hundred, contribute to human aging.

A more formidable number was posited in 1978 by University of Washington gerontologist George M. Martin. Based on a sweeping review of genetic diseases, he estimated that up to seven thousand "genetic loci" are probably involved in controlling aging, though he added that perhaps as few as seventy control aging's "major" aspects. He concluded, "It is naive to believe that a mutation at a single [gene] locus could be responsible for the determination of life span and the various debilities of aging."

Those who dreamed of finding genetic simplicity at the heart of aging—and perhaps even single genes with profound effects on life span—had a compelling practical reason to entertain their seemingly naïve hope. Drug developers are perfectly capable of developing medicines that counteract the effects of single genes. (To be precise, most drugs disable proteins that genes make, rather than the genes themselves. For simplicity, let's assume for now that genes are the targets.) But attempting to pharmaceutically alter a throng of mostly random, largely independent processes in the body, each regulated by multiple genes, would bring to mind, once again, the little Dutch boy trying to plug thousands of dike leaks. Moreover, engineering a drug that targets many genes at once, if you'll permit another simile, is like trying to design a single key that can open multiple locks of different sizes and designs. Any sensible pharmaceutical expert would consign such a challenge to the "limbo of scientific impossibilities," to borrow Williams's phrase.

You might counter that developing a combination therapy containing multiple drugs, each aimed at a separate gene, could do the trick. But the risk, complexity, and expense would likely be prohibitive.

Thus, anti-aging optimists hoped that even if many genes impinge on aging, they would be organized like an army: a top-down hierarchy with a small number of high-ranking genes that, if given the right pharmaceutical order, would mobilize many subsidiary genes to throw their weight against the aging process. The chance of that being the case seemed very remote to the pessimists, and from their perspective the optimists were arguably naïve, possibly con artists—or maybe just mad as hatters.

But like Lord Kelvin, Edison, and Einstein, the skeptics were in for a great surprise. By the late 1970s, experiments were under way that would show that a mutation in just one gene could actually double an animal's life span. The finding would begin a transformation in gerontology akin to the one triggered in medicine by the revelation that infectious microbes are behind many of humankind's worst scourges. That earlier advance, it's worth recalling, made it possible to avert untold suffering and death that had once seemed inevitable.

. . .

If you walk across a patch of ground almost anywhere in the earth's temperate zones, you're likely to tread on nematodes, the ubiquitous set of worms that includes *C. elegans*. You don't necessarily have to go walking to get close to them, though. In his entertaining history of nematode research, *In the Beginning Was the Worm*, science writer Andrew Brown notes that "there is a gruesome saying among worm researchers that if everything on earth were to disappear except the nematodes, the outline of all plants and animals would be left, filled out by their nematode parasites."

The worm, as *C. elegans* is known in science, may not be the most widespread nematode, but worm fanciers have found it, among other places, in Australia, England, France, Germany, Hawaii, and all across the continental United States. When Sydney Brenner, the Nobel laureate who popularized *C. elegans* for research, decided in the 1960s to investigate its suitability as a model for analyzing the development of the nervous system, he simply strolled into his garden in Cambridge, England, and collected a bevy of worms from the soil—the specimens would later be known as the N1 strain. The main *C. elegans* strain he disseminated to other researchers, however, originated in a compost heap near Bristol, England. Dubbed N2, they came to him by way of Ellsworth Dougherty, a biologist at the University of California at Berkeley who had a knack for collecting stuff that would later be seen as priceless treasure. Besides channeling superlative worms to posterity, Dougherty assembled a fabulous collection of Wizard of Oz memorabilia.

Scientists believe that *C. elegans* owes its ubiquity largely to a talent for stretching its life span like a rubber band. The worms can sense when they are threatened with famine due to a shrinking supply of the bacteria they feed on, or from a population explosion of fellow worms in their vicinity. Instead of merrily feasting themselves into oblivion when this happens, they hunker down in the "dauer" stage, a prolonged larval phase designated by the German word for enduring.

Dauer-bound, they sock away extra fat like tiny camels and encase themselves in tough, water-repelling sheaths. Though still capable of wriggling toward food and water, dauer worms mainly wait for good times to return, when it will make sense to develop into adults and reproduce. They can endure nearly three months in this sporelike state—in times of plenty they live less than three weeks. Their ability to wait out lean times enables them to survive in a wide array of iffy niches.

C. elegans' telescoping life span has made the worms especially intriguing to gerontologists, and in 1975, the newly formed National Institute on Aging announced plans to fund worm-based research on aging. A year later, one of the first worm studies funded by the NIA appeared in *Nature*, which at the time rarely deemed gerontologists' work up to its standards. The report, by two young researchers at the University of Colorado, David Hirsh and his postdoctoral protégé, Michael Klass, showed that worms in the dauer state don't, as suspected, age more slowly than usual—they essentially stop aging altogether. The scientists found that after spending the equivalent of several normal lifetimes as dauers, worms go on to have adult lives that are just as long as those of worms that never enter dormancy. That meant that during their dauer days, they manage to suspend the progression of free radical and other damage behind aging. Fascinated by this time-out mechanism, Klass spent the next decade trying to figure out how it works.

He began by assuming that the dauer stage, including its resistance to aging, is genetically regulated. That was a reasonable premise given that the dauer state is basically just a step, albeit a remarkably long one, in the development of a worm from larva to adult, and it was becoming clear at the time that all such early development unfurls under tight genetic control. Thus, Klass theorized that by mutating worm genes, he might be able to switch on the animal's anti-aging mechanisms, possibly without also inducing full-tilt dormancy. The idea was to isolate life-extending genes that might have analogs in higher animals, like us.

It was a fascinating pursuit, but very daunting. For one thing, Klass would have to mutate genes with a chemical that acts like a shotgun fired at DNA, and it might well take many years of blasting away to induce mutations in key aging-related genes. To find out whether he had hit such genes, he would have to monitor many thousands of mutated nematodes from birth to death to measure their life spans, a tedious process that entails repeatedly prodding old nematodes to see whether it gets a rise out of them. (If it doesn't, they're goners.) And even if he got lucky and found abnormally long-lived mutants, he wouldn't necessarily be able to identify the genes behind their longevity. The main gene-mapping technique at the time involved laboriously tracking the inheritance pattern of a trait of interest in order to get a rough idea of where the gene responsible for it lay in the genome, and it wasn't very good at pinpointing genes of interest.*

No wonder few young molecular biologists opted to go into gerontology in those days. Klass was thoroughly hooked, though. For years, he told me in an interview, "I mutated worms and then looked and looked and looked" for ones that lived longer than usual. "It was brute force. I had two, three, four thousand plates [of worms] going in the lab. People would look at me and say, 'Ohhh-kay.'"

After landing a faculty post at the University of Houston, he continued doing much of the grunt work himself—he didn't have much luck attracting graduate students to lend a hand. The payoff, when it finally came, was disappointing. In 1983, he reported the discovery of five strains of long-lived mutant worms that at first glance looked promising. But on close inspection, they turned out to carry a mutation that caused lethargy and appeared to limit their ability to ingest food. Klass surmised that their longevity was due to calorie restriction—the well-known, slow-aging phenomenon induced by low calorie intake—rather than to novel anti-aging genes. Sounding

---

*Genes located near one another on DNA molecules tend to be inherited together. Thus, if abnormally long life in worms were usually inherited along with a trait whose gene location is known, say, infertility, scientists would know that the long-life gene, or genes, is near the fertility-related one. By repeatedly using this fellow-traveler principle with previously mapped genes, they could get an increasingly precise fix on the long-life gene's location in the genome.

dispirited in his report, he concluded that "these results appear to indicate that specific life span genes are extremely rare."

Three years later, Klass gave up aging research and left academia to join Abbott Labs, the health-care company based near Chicago. "After more than ten years studying worm aging, I was ready to do something else," he said. "My marriage had fallen apart, and I wanted to move closer to my family in Wisconsin. I also wanted to be able to tell my mother what I worked on and have her say, 'That's cool.'"

· · ·

Klass's long-lived worms may have lacked mom-wowing power, but their research careers weren't over. After Klass put them aside, he sent a frozen batch of them to a colleague, Thomas E. Johnson. (Nematodes can be stored in liquid nitrogen and revived years later.) Like Klass, Johnson had gotten interested in worm aging as a postdoc at the University of Colorado, where the two had met in the late 1970s. In 1982, Johnson was named as an assistant professor at the University of California at Irvine. Casting about for a major research project, the young prof decided to thaw Klass's worms and take another look at them. He soon noticed that although they were sluggish and lived longer than usual, as Klass had reported, they didn't seem to eat less than normal worms do. That meant that they probably weren't living longer merely as a result of calorie restriction—and that Klass may have actually found what he'd set out to discover without realizing it.

Still, Johnson didn't expect to discover an anti-aging gene that Klass had missed. Like most gerontologists, he thought it was extremely unlikely that a single gene could have a large effect on aging. In fact, he hoped to confirm the conventional wisdom by showing that many aging-related genes had been mutated in Klass's worms, and that each had made a small contribution to their elongated life spans. The very idea that just one gene was behind their dramatically lengthened lives "sort of stuck in my craw," he told me in an interview. After all, he knew that a single exposure of a worm to the gene-busting chemical Klass had used could induce some twenty mutations across a nematode's genome at one fell swoop.

To winnow out possible anti-aging genes, Johnson employed the same laborious strategy Klass had used—he crossed the long-lived mutants with other strains of worms, then analyzed how the long-life trait was inherited. He had help. An undergraduate yearning to do hands-on research, David Friedman, volunteered to work in his lab after attending one of Johnson's first lectures as a biology teacher—Friedman was a freshman at the time and hadn't learned to look down on aging research. With his lab short-handed, Johnson was perfectly happy to give the enthusiastic youngster a crash course in worm science. In return, Friedman served as his right-hand man for more than four years.

They soon discovered an oddity: Crossing the long-lived mutants with unmutated worms yielded offspring with normal life spans. The simplest explanation seemed unbelievable at first—one could account for what had happened by presuming that the mutants' longevity had been conferred by a single recessive gene whose anti-aging effect was masked when it was combined in the offspring with a dominant counterpart that causes normal life span. (Recall from Bio 101 that genes are generally found in cells in the form of paired variants, called alleles, and that dominant alleles override the effects of recessive ones.) But as the two did more crosses, the results insistently argued that they had indeed, as mind-blowing as it seemed, found a lone gene that dramatically extends life span.

Suddenly Johnson found himself parting ways with most of his field's senior scientists, some of whom would express doubts about his finding for years to come. But he was the most equable and disarming of firebrands. A burly, balding man with a fringe of red hair, he had been distancing himself from his elders for almost as long as he could remember.

His father, a Denver ironworker who never completed grade school, was the youngest of seventeen kids and had left home at age twelve to make his own way after both of his parents had died. "My dad had an incredibly hard life, and he didn't think it made much sense for me to go on to high school, let alone college," Johnson recalled. "My mom never graduated from high school either"—she was only seventeen when Johnson was born. "But she saw higher education as a big thing and was pretty supportive of me." From an early age, he made it clear that he wasn't cut out for the blue-collar life. "I was always running little experiments, often to the detriment of my mom's things," he recalled. "I remember at about age three watching a hairbrush she'd gotten as a wedding present melting in the oven, because I was curious to see how fast it would drip down to the bottom."

Impressed by his drive and smarts, his mentors at Regis, the Jesuit-run high school in Aurora, Colorado, that he attended, encouraged him to apply to the Massachusetts Institute of Technology. When the prestigious school accepted him, the Cinderella story of his youth was complete. The ironworker's son is now a professor at the University of Colorado and recently served as president of the American Aging Association, a leading supporter of gerontology research.

After concluding that he really had discovered the first "gerontogene," a term he didn't coin but popularized, Johnson dubbed it age-1 and published the finding in 1988. When the paper was accepted by the journal *Genetics*, he and Friedman, named as coauthor, broke out a bottle of champagne that had long been part of the lab's unused apparatus. Among other things, they had discovered that age-1 could more than double worms' maximum life spans. But they hadn't been able to isolate the age-1 gene, which meant they couldn't decode its sequence of chemical units, nor discover its basic function in normal worms—steps necessary to find out whether a similar gene exists in mammals.

With so much about age-1 still shrouded in mystery, Johnson had trouble convincing peer reviewers that his next major finding on it was sound enough to publish—he'd demonstrated that age-1 met the gold standard for anti-aging effects by slowing worms' normal rise in mortality risk as they get older. One reviewer, a prominent nematode researcher who is now deceased, insisted that Johnson send him the long-lived mutants so that he could replicate the study in his own lab before he would approve its publication, Johnson told me. It was an insulting, unheard-of requirement, and it delayed the report's appearance for eighteen months. The study finally appeared in *Science* in 1990. When I asked Johnson how he dealt with the frustration, he replied, "One of my problems is that I never figured out when I had the right to get mad."

· · ·

Tom Johnson's worms made some waves, but the existence of geronto-genes was neither well accepted within gerontology nor widely recognized outside it until the field's first media star put her imprimatur on the story. Cynthia Kenyon had already established herself as a leading nematode geneticist when she took up the quest for life-extending *C. elegans* genes in the early 1990s. In 1993, her lab at the University of California at San Francisco struck gold by discovering that a mutated gene called daf-2 could more than double worms' life spans.

Importantly, the gene's function had long been known to nematode researchers. Daf-2 is one of a group of dauer formation genes that, when mutated, abnormally induce the state of dormancy in larval worms. Discovered a decade earlier by Donald Riddle, now at the University of British Columbia in Vancouver, daf-2 mutations essentially make worms hallucinate an impending famine and go into the dauer state even when food is plentiful. What made Kenyon famous, in a nutshell, is that she demonstrated that the mutations could also extend life span in worms that never enter the dormant state.

Scientists' familiarity with dauer worms' long lives made them more willing to buy Kenyon's discovery than Johnson's. Indeed, when she presented the discovery at a meeting of worm scientists several months before it was published, there was little excitement—most saw it as a minor extension of what they already knew about daf-2.

Before long, though, the implications sank in and caused a stir. By growing her daf-2 mutants at cool temperatures during their early development, Kenyon had been able to prevent their mutation from taking effect and causing them to enter dormancy. Later, when they had matured beyond the larval stage at which dauer formation can be triggered, she raised the temperature, presumably switching on daf-2's anti-aging effects for the rest of their lives. The fact that they lived more than twice as long as usual—all the while appearing essentially normal (they did show a slight decrease in fertility)—demonstrated that daf-2's magic could be uncoupled from the dauer state. That in turn suggested that the power of a single gene to retard aging might not be limited to the peculiar little corner of the universe occupied by worms that stretch their life spans by going into suspended animation.

Kenyon's report boldly concluded that figuring out how daf-2 does its anti-aging thing "could lead to a general understanding of how lifespan can be extended." And as the report made clear, her lab was already striving to do just that by unraveling the daf-2 pathway, the set of interacting genes and proteins that spread daf-2's signal through worms' cells to shift the creatures into slow-aging mode. Drawing on earlier studies by Riddle and other worm researchers, she had quickly zeroed in on a gene called daf-16, showing that it was a key relay in *C. elegans*' anti-aging machine—mutating daf-2 activated daf-16, which then switched on a panoply of other life-span-extending genes.

Unlike Johnson, Kenyon had no trouble placing her report in a prominent venue. *Nature* published it in December 1993, along with an arrestingly titled commentary, "Methuselah among nematodes." For the first time, the possibility of intervening in the aging process began to seem real to a lot of people.

Kenyon soon eclipsed Johnson as the media's go-to expert on geronto-genes. A slim, vivacious, blond-haired woman who looks as if she's ben-efited from spending time in a dauer phase herself, she enthuses about her work in a punchy, lucid way that's rare in science. As if that weren't enough to make her a media favorite, she plays a big alpine horn in her spare time, spices her explanations with references to Shakespeare's sonnets, and keeps a copy of Lewis Carroll's *Alice's Adventures in Wonder-land* on her office bookshelf. (In my view, Carroll's works should be part of every gerontologist's library, if only because of his instructive passage about the White Queen teaching herself to believe up to six impossible things before breakfast.) After experiencing the full force of Kenyon's magnetic field during an episode of *Scientific American Frontiers* that PBS aired in 2000, host Alan Alda sounded as if he were ready to award her a Nobel Prize on the spot. (For years she has been mentioned in the press as a contender.) "Something like a long-life pill could be possible once we fully understand" anti-aging genes, he ventured. "And someday we surely will, thanks to Cynthia Kenyon."

Most gerontologists tend to fall back on cautionary clichés when talking to the press about the possibility that their research might advance the hunt for anti-aging drugs. But Kenyon has routinely tossed out zingers like "fountain of youth gene" to grateful reporters who have made pilgrimages to her lab over the years. With her distinguished track record, she can afford to say things that would give a less prominent scientist a reputation as a hyperbolic lightweight. When Alda asked her on the air whether the advances she was spearheading implied that he could rightly entertain the hope of living to 140, she shot back, "Why not?"

A bookish polymath as a youngster, Kenyon was known in her family for pursuing girl-scientist pastimes, such as keeping a praying mantis on a string and feeding it with a honey-dipped toothpick. In 1976, she graduated as class valedictorian in chemistry and biochemistry at the University of Georgia, where her father was a geography professor and her mother an administrator in the physics department. She went to the Massachusetts Institute of Technology for her Ph.D., then on to Britain's Laboratory of Molecular Biology in Cambridge, where she trained under Sydney Brenner, the father of *C. elegans* gene research. Before getting interested in aging, she established herself as one of Brenner's star protégés by elucidating genes that help orchestrate the formation of nematodes' and other organisms' bodies. In 1986 she joined the faculty at UCSF, where she is now a professor and director of the Hillblom Center for the Biology of Aging.

Besides making a series of key contributions to the genetics of aging, Kenyon deserves credit for drawing unprecedented public attention to the research and its implications. But her tendency to upstage peers is considered by some other worm researchers to violate the informal code of collegiality that they regard as setting their tribe apart from others in biology. When journalists have written about her 1993 discovery, they've often portrayed it as utterly unprecedented. The error isn't entirely the media's fault—Kenyon's much-cited report on daf-2 didn't mention Johnson's earlier discovery of a single gene mutation that can double *C. elegans'* life span.*

Still, progress in science, as in other pursuits, is carried along on

---

*Once I asked Johnson whether he resented the fact that his pioneering work on anti-aging genes has received far less notice than Kenyon's has. It didn't seem to bother him much. In fact, he mildly observed, the validity of his age-1 discovery was widely doubted until Kenyon reported her similar finding, and thus he had felt more buoyed than deflated by the buzz about daf-2.

the flow of competitive juices. And Kenyon has helped motivate rivals to press ahead with the same intensity she has shown in the race to find and elucidate new life-extending genes—an intensity that suggests gerontology's first Nobel Prize hangs in the balance.

Ironically, though, a round of major discoveries about worm gerontogenes immediately following the daf-2 finding was led by a scientist who had no desire to study aging and dislikes the one-upmanship needed to win respect and funding in science. We should all be grateful that Gary Ruvkun can't always get what he wants.

. . .

When Kenyon identified daf-2 as a gerontogene, Ruvkun, whose lab had been studying the gene for reasons unrelated to aging, was nonplussed. "I thought, Oh gosh, now I'm in aging research. Your IQ halves every year you're in it," he told me in a 1999 interview.

His comic hyperbole wasn't aimed at fellow scientists. Like many biologists at the time, he thought the study of aging was dominated by self-anointed experts hawking miracle elixirs. But Ruvkun, a geneticist at Boston's Massachusetts General Hospital and professor at Harvard Medical School, soon found himself swept up in serious aging science. In fact, he'd already contributed to the field without realizing it. With Shoshanna Gottlieb, a postdoc in his lab, he had found evidence that daf-2's dauer-related effects are funneled through daf-16. They publicized the finding in mid-1992 in the *Worm Breeder's Gazette*, an informal outlet, now defunct, in which nematode researchers often published briefs on findings that hadn't yet appeared in a journal. Thus, Ruvkun had reported how daf-2 and daf-16 work together before Kenyon revealed their involvement in aging.

In 1994 Ruvkun continued his unplanned adventure in gerontology. He and Gottlieb reported that a gene called daf-23 works a lot like daf-2 to trigger the dauer state, and, like daf-2, exerts its effects via daf-16. Soon after, nematode researcher James Thomas at the University of Washington discovered that daf-23 and Tom Johnson's age-1 are the same gene. Once again Ruvkun had spearheaded an important discovery on aging without meaning to.

Yielding to fate, around 1995 he hung out his shingle as a part-time gerontologist by launching a series of seminal studies on gerontogenes—today he even keeps a totemic can of Longevity Brand Sweetened Condensed Milk in his office. (His main claim to fame, though, is his work outside gerontology—over the past decade, he has helped open a new chapter in genetics by showing that small strands of RNA, DNA-like molecules long regarded as supporting actors in cells, actually play leading roles in the regulation of genes. In 2008 this "microRNA" research earned him a Lasker Award, a distinction that often presages a Nobel Prize.)

A tall, droll, avuncular man resembling Gene Shalit with a normal mustache, Ruvkun is known for leaping with such abandon from one provocative possibility to another during his lab's brainstorming sessions that his young protégés often find themselves playing the conservative grown-up, trying to bring their mentor back to earth. A standing joke in the lab is that Ruvkun is willing to bet three dollars on each of his hunches, no matter how iffy, and in theory now owes his collected grad students a gob of money. When describing him, they use terms such as "warmhearted" and "supportive" with notable consistency. "If we didn't understand the results of an experiment, we could go into his office and he would talk to us about it for four hours," said Heidi Tissenbaum, a former postdoc in Ruvkun's lab who is now an associate professor at the University of Massachusetts Medical School in Worcester. "He gets excited about things."

Ruvkun might be described as a serial enthusiast who, paradoxically, has a lot of stick-to-itiveness. His passions, in rough chronological order, have included amateur astronomy, electrical engineering (growing up in the Oakland area, he wanted to be an engineer, like his dad), physics (he fell in love with it at college), medicine (after getting a biophysics degree at the University of California at Berkeley, he applied to med school but hadn't taken required courses and was rejected), the ramblin'-man life (he went on the road for two years in the 1970s, first roaming around the Pacific Northwest while living out of a Dodge van, then meandering through South America on third-class buses as far south as Tierra del Fuego), having enough money to eat (after rambling, he worked for a year as a medical technician to make ends meet), a girlfriend in Boston (to be near her, he applied to Harvard's biology graduate program and got in), worm genetics (after getting his Ph.D., he worked as a postdoc in the MIT lab of Robert Horvitz, who shared a 2002 Nobel Prize for gene discoveries in nematodes), and life on other planets (in recent years he's helped develop microbe detectors to send to Mars). "I go off on tangents," he told me.

After Ruvkun got seriously interested in aging, he quickly emerged as one of gerontology's leading gene finders. In 1996, his lab sifted Johnson's mysterious age-1 from the worm genome. It wasn't a very informative finding, though. The gene turned out to engender part of a "kinase" enzyme, which said little more than that it helps regulate cells' internal operations.

But the isolation of daf-2 and daf-16 still beckoned. Biologists knew the two anti-aging genes existed somewhere in the worm genome, but not enough was known about them to get a firm handle on their implications for human medicine.

Ruvkun's team soon narrowed daf-2's approximate location to a segment of DNA spanning multiple genes. When scanning the segment for clues to daf-2's identity, they noticed a gene snippet resembling part of the gene for a well-known hormone receptor—the one that conveys the signal from insulin, the blood sugar regulator, into mammalian cells. That clicked: Given daf-2's hormonelike ability to trigger body-wide change, Ruvkun had a hunch that its gene might well be a component of a hormone system. "I thought, This is perfect. This has to be it," he recounted. It was.

Further analysis showed that the gene engenders an antique version of two mammalian hormone "receptors," molecules that protrude from cells like receiving antennae for hormonally conveyed signals. One conveys signals from insulin into mammals' cells. The other transmits signals from a structurally similar hormone called IGF-1 (insulin-like growth factor 1), which is thought mainly to stimulate growth.

It was a truly mom-wowing gerontogene discovery: Ruvkun's lab had shown that a worm anti-aging gene has human counterparts. What's more, the counterparts engender proteins that conveniently stick out of cells, making them readily accessible to drug molecules. The group was soon poring over the medical literature on the two hormones, searching for evidence that fiddling with them has effects on aging.

Intriguingly, some humans carry mutations that knock out their insulin receptors, a situation somewhat similar to mutating daf-2 in worms. Unfortunately, instead of living longer, such people suffer from leprechaunism, a rare syndrome that severely retards growth and usually is fatal during early childhood. When Ruvkun asked endocrinologists he knew whether they'd ever seen an adult leprechaun—which would indicate that such mutations aren't necessarily harmful—they said no.

But he suspected that mutations that only partially disable the insulin receptor might have very different effects. Indeed, such "weak" mutations of daf-2 appear to be the kind that extend life span in worms. There seemed a chance that similar mutations in people might induce metabolic changes akin to the fat-storing, hunkered-down dauer phase in worms—something like what happens to bears as they get ready to hibernate. That possibility seemed especially plausible after the researchers unearthed an obscure Japanese study on a human patient, a teenager who carried an insulin-receptor mutation that appeared to mimic a weak daf-2 mutation. The fourteen-year-old was reportedly morbidly obese and diabetic.

Pursuing a characteristically provocative line of thought, Ruvkun speculates that such gene variants may have evolved in humans to help them outlast famines and are now widespread in the population. If so, such genes might partly explain skyrocketing rates of obesity and diabetes—somehow our version of worms' fatten-up-hunker-down mode is kicking in inappropriately at a time when rich food just keeps coming at us, shortening rather than lengthening our lives.

After nailing daf-2, Ruvkun focused on daf-16, the key relay through which the anti-aging effects of daf-2 and age-1 are channeled. By summer 1997, his group had pinpointed the gene but hadn't yet submitted a paper on it. Meanwhile, Kenyon's lab was also scrambling to pin it down. The record would later show that it was a close race. Ruvkun's daf-16 paper reached *Nature* on August 20, one day after Kenyon's paper was received at *Science*. But in the end the Ruvkun lab's paper, which included a more detailed analysis of what daf-16 does than did the rival report, was published first.

Daf-16 turned out to be the gene for a "fork head" transcription factor, a pronged protein that interacts with DNA to activate various other genes. It might be described as the Rome of aging genes—many other mutations discovered to extend worms' life spans have been found to exert their effects via daf-16, making it seem like the hub of an ancient civilization whose reach is being mapped out by the delineation of major trade routes that go through it.

After worm scientists proved the existence of gerontogenes, fruit fly researchers began scanning their own favorite animal for similar genes and soon found one. In 1998, a giant of fly research, the late Seymour Benzer of the California Institute of Technology, reported the discovery of a gene, dubbed "methuselah," that when mutated extends flies' average life spans by 35 percent. Intriguingly, methuselah flies were found to weigh about one-fourth more than normal ones—as if they'd stored up fat to outlast a famine.

• • •

You might think that the discovery of gerontogenes would have quickly changed everything in gerontology. But some of the field's leading figures long continued to downplay, or simply to ignore, the implications of the advance. In 2004, for example, one of its senior statesmen, Leonard Hayflick, wrote that because of the "random downward spiral of molecular disorder . . . that we call aging," it's "doubtful that intervention in the aging process has been achieved in any . . . life form." Further, the likelihood of doing so "is remote." And for those who think they're making headway toward anti-aging drugs, he added, "The practice of 'anti-aging medicine' is the second oldest profession and it shares much with the oldest."

To most gerontologists, however, the discovery of gerontogenes seemed both real and important, as well as a kind of comeuppance for evolutionary biologists, who had been the leading skeptics about the existence of such genes. Indeed, one of the subtexts of the geron-togene story is that living things sometimes aren't as predictable as Darwinian theorists think.

Still, the overarching evolutionary theory of aging was never threatened. Worms' gerontogenes can be seen as elements of a genetic module that evolved to boost fitness by enabling the animals to take a time-out early in life when things get rough, boosting their chance of passing genes to future generations. And as Kenyon showed, muta-tions in the genes can abnormally switch on their anti-aging effect in adults. The mistake of theorists like Williams lay in failing to antici-pate this surprising offshoot of their general theory, not in getting the theory wrong. Here's another thing they missed: Like other gene modules that come into play during development, the anti-aging one is hierarchically organized. That explains why just one gene can exert such a shockingly large effect on aging—gerontogenes like daf-2 are near the top of the hierarchy, and thus they coordinate the action of scores of subsidiary genes to slow aging.

But Kenyon's bold optimism notwithstanding, in the mid-1990s few scientists were convinced that studying gerontogenes was likely to lead to breakthroughs in human medicine. After all, although we carry genes similar to ones that evolution employed when crafting worms' dauer module, very few of us turn into living mummies when we get hungry or find ourselves hemmed in by a crowd. There was little reason to believe that our genomes include a dauerlike module that's just begging to be activated with anti-aging drugs.

Even if we possessed such a module, we might pay a heavy price for tampering with it. Indeed, it seemed likely to many scientists that long-lived mutant worms harbor subtle defects that don't show their effects inside the cuddly environs of the lab. Otherwise, why wouldn't evolution have permanently activated worms' anti-aging machine, lengthening the animals' reproductive spans? Adding sub-stance to this idea, a 2000 study led by Gordon Lithgow, a geron-tologist at the Buck Institute in California, showed that when age-1 and normal worms were both subjected to a fluctuating food sup-ply, which is thought to mimic natural conditions, the long-lived mutants rapidly died out while the normal ones didn't. Long-lived mutant worms, it appears, are secret softies that can't hack it in the real world.

. . .

Despite doubts that gerontogene research would yield medical advances, many scientists jumped into the race to unravel how they work. Jacques Vanfleteren at the University of Ghent in Belgium and Pamela Larsen at MIT were among the first to shed light on that—they independently discovered that the age-1 mutation boosts worms' antioxidant defenses, enabling them to laugh off free radical hits that kill normal worms. Tom Johnson, who'd returned to the University of Colorado, followed up by showing that age-1 worms are resistant to heat stress. As signs of long-lived worms' resistance to hurtful things accumulated, he connected the dots to come up with a compelling hypothesis on what life-extending genes are really up to.

Johnson theorized that genes like daf-2 mainly turn on various kinds of stress-resistance pathways, hardening worms against a variety of threats they face in the cruel world while in slow-aging mode. And even when they're not hunkered down in the dauer state, stresses they encounter during the normal course of existence, such as exposure to high temperatures or chemicals that release free radicals, can switch on their stress response—the anti-aging module apparently comes in handy to help them cope with a variety of dicey situations. In fact, Johnson and colleagues showed that when normal adult worms are briefly exposed to high temperatures, their lives are extended by about 15 percent. This toughening-up phenomenon, called hormesis, brings to mind Nietzsche's pithy observation that "what doesn't kill us makes us stronger."

The close tie between the stress response and durability isn't just a worm thing. The mutant methuselah flies discovered by Benzer's lab, for example, also are hardened against various stresses. In light of other such findings, it now seems likely that virtually every organism possesses a version of the stress-triggered anti-aging module. But over the eons, the blind watchmaker appears to have customized both the module's "front end"—the particular pathways that turn on the module—and its outputs in different creatures to reflect the particular stresses each tends to encounter.

For example, the worm version is geared to start up, along with the rest of the dauer machinery, when a worm detects high concentrations of a pheromone emitted by its fellows, which indicates to the animal that its kind are getting so thick on the ground that food will soon run short. Intriguingly, Kenyon and colleagues have shown that knocking out worms' smell and taste neurons can also switch on the module and extend the animals' life spans—loss of the ability to sense food apparently fools them into acting as if famine is coming.

As for the module's anti-aging outputs in worms, it appears that an important one involves the ramping up of antioxidant enzymes. That makes sense, because it's likely that short-lived worms normally devote little of their inner resources to blocking free radical damage, and therefore would have to put special emphasis on doing so to maintain themselves in a youthful state.

Irrespective of such customization, the module's function and many of its core genes appear to have remained largely intact for more than a billion years, enabling animals to harden themselves against stresses that threaten to cut them off before they reproduce. And that gave reason to hope that anti-aging genes exist in humans despite the fact that we mammals have no dauer phase.

Inspired by that insight, Johnson cofounded a biotech company in 1997 to develop drugs based on his gerontogene research. Named GenoPlex, the start-up planned to pursue the strategy pioneered by Eukarion—it would exploit anti-aging research to develop orthodox medicines for diseases of aging. But GenoPlex lasted only three years before folding, a victim of oversized ambitions and undersized funding. Setting out with only $1.2 million of seed money from venture capitalists and a mutual fund, the start-up had hoped to develop better anesthetics, Alzheimer's disease therapies, and, in time, an array of gerontogene-based medicines.

But even if GenoPlex had raised a ton of money, it's not clear that it would have gotten very far trying to turn worm gerontogene research into human drugs. As we saw earlier, daf-2-like mutations in humans have profoundly adverse effects, so pursuing medicines to emulate them would probably be a bad idea. Mimicking other worm geronto-genes with drugs might have similar downsides.

Arguably, what scientists needed in order to make real progress in the quest for anti-aging drugs were targets closer to home—mammalian gerontogenes. But did they really exist? The answer, like many other seminal advances in gerontology, came from an unexpected quarter. Carbondale, Illinois, to be exact.

5

THE REALLY STRANGE THING ABOUT DWARFS

CENTURIES BEFORE WALT Disney got rich with Mickey, the entrepreneurs of the East found a way to turn mice into gold. In 1654, a Buddhist priest named In-gen traveled from China to Japan with a pair of striking white mice with jet-black eyes. Their rare beauty captivated his acolytes, one of whom bred the mice as collectibles, launching a fad that let him attain financial nirvana. A century later a guild of meticulous rodent breeders had formed in Japan. They crafted a dazzling array of jewels for Edo period mouse fanciers: mice with lilac-colored heads; black mice whose breasts were marked with white crescents; the rare dark pink "Azami" mice; and, most cunning of all, "Mame-nezumi," dwarf mice no bigger than a man's thumb. This chapter tells the story of these dwarfs, which were more remarkable than even the most adoring mouse fancier realized. Long regarded as delicate flowers, they were actually supermice, harboring a stunning longevity secret that wasn't discovered until the 1990s.

It's unclear whether dwarfing genes passed from the tiny eighteenth-century Japanese mice to dwarf mice used in research today. One reason is that spontaneous mutations in four different genes are known to produce dwarf mice, and such mutations have arisen more than once in breeders' colonies over the centuries. Still, it's possible that growth-retarding genes carried by Mame-nezumi mice made their way from Japan to Europe during the late nineteenth century. That's when a rage for "fancy" mice took hold in Europe, prompting Victorian fanciers to import the East's furry treasures along with its tea, porcelains, and silk. Around 1900, American aficionados began importing lots of fancy mice, and some years later the dwarfs turned up in the United States.

Not long after the West's fancy-mouse fad took off, the pioneers of genetics began casting about for a mammal to use in their efforts to link readily identifiable traits to genes. At a fateful meeting around 1907, Harvard zoologist William Castle, one of America's first big-name geneticists, skidded a live mouse across a bench top to one of his students, Clarence Cook Little, and told him to learn all about it. Presumably, Little had quick hands—there's nothing in the record to suggest that there was a frantic mouse chase that day at Harvard. We don't have to guess about the quickness of Little's wit, though. An academic standout and captain of Harvard's track team, Castle's "mouse man," as he soon became known, possessed the energy, drive, and craftsmanship of his great-great-grandfather Paul Revere. Over the next few years he spearheaded the transformation of the bane of the granary into the boon of the laboratory. Little developed the first genetically uniform, inbred strains of mice, including ones still widely used in research, and in 1929 he founded the Jackson Laboratory, a world-leading center for mouse genetics, in Bar Harbor, Maine.

Showing that behind every great mouse man there's a woman with a thing for mice, Castle's group obtained most of their original mice from a former schoolmarm named Abbie E. C. Lathrop. Forced to retire from teaching at age thirty-two by pernicious anemia, in 1900 she took up poultry farming in the Massachusetts hamlet of Granby. But the clever Miss Lathrop soon switched to a more lucrative business: breeding fancy mice. She started with a single pair of Japanese waltzers, so named because of their curious, dancelike movements. (One of Castle's students later discovered that their strange behavior was due to inner-ear defects.) Before long, she'd become a mouse rancher extraordinaire. By 1913 she'd filled her barn and sheds with more than ten thousand mice kept in straw-filled boxes, and geneticists from all over were ordering Granby fancies. Lathrop even helped carry out early mouse studies on the genetics of cancer.

After Little left Harvard, Castle brought in another extremely able mouse man: George D. Snell, who would later win a Nobel Prize for mouse-based discoveries about the immune system. In late 1929, almost exactly a year after Mickey Mouse made his debut in the animated short *Steamboat Willie*, Snell reported that some odd little mice had turned up in Harvard's colony. He soon discovered that the stunted, snub-nosed rodents carried a recessive gene that completely arrested growth at about two weeks of age.

The Snell dwarf, as it is now called, appeared to be a feeble midget. At maturity, it was only about a third the size of a normal mouse and, in Snell's words, had "sub-normal" vigor, "showing very little tendency to run about and never trying to jump out of the jar" in which it was placed for examination. He thought the dwarfs would be excellent for studying growth-related genes. But there were so many mice, so little time: Snell moved on, leaving others to unravel the dwarfs' genetic defect.

By the 1950s, scientists had identified the primary effect of the Snell dwarf's stunting gene: It blocked the formation of part of the pituitary gland, causing a profound deficiency of hormones that drive bodily development, including the main one, growth hormone, as well as thyroid stimulating hormone, which helps regulate energy metabolism via the thyroid gland. Many scientists suspected that the dwarfs' pituitary defect shortened their lives as well as their bodies. In fact, it seemed so obvious that the mice were short-lived that no one bothered to rigorously test the idea. You'd expect puny, lethargic rodents with a profound endocrine deficiency to die young, wouldn't you? So why go to all the trouble of conducting a laborious life-span study on them?

Besides, anecdotal evidence suggested that human dwarfs with similar pituitary deficiencies are often short-lived. The most famous one of all, Charles Sherwood Stratton, who, as General Tom Thumb, served as one of P. T. Barnum's greatest attractions during the heyday of the nineteenth-century circus, had died at forty-five. His 1883 obituary in the *New York Times* noted that his life "had been a rather long one" for a dwarf. Another celebrated dwarf in Barnum's circus, George Washington Morrison "Commodore" Nutt, had died two years earlier of kidney disease in his late thirties.

In 1972, a study published in *Nature* seemed to remove whatever doubt remained about the life-shortening effect of dwarfing genes. Its data showed that Snell dwarfs not only die young—at about four and a half months, versus the average life span of twenty months for normal mice—but also experience signs of accelerated aging, such as prematurely discolored hair, balding, and cataracts. Led by immunologist Nicola Fabris at the University of Pavia in Italy, the report suggested that the root cause of the dwarfs' "precocious" aging was an arrested development of their immune systems. It was an appealing hypothesis, meshing perfectly with everything scientists knew about dwarfs' apparently petite life spans. The only problem was that it was dead wrong.

. . .

It's often said that the thing that most sets science apart from—and above—earlier modes of inquiry is its self-correcting nature. That may be so. But sometimes the self-correction operates with all the speed and efficiency of early experimental aircraft propelled by flapping wings. During the two decades following publication of Fabris's study on dwarf mice, multiple reports appeared that baldly contradicted its conclusion about the brevity of their life spans. The conflicting data had no noticeable effect on broad biomedical opinion.

The first blow against the conventional wisdom was delivered the same year Fabris's report appeared: Ruth Silberberg, a painstaking, German-born pathologist at Washington University in St. Louis, observed that Snell dwarfs have "unusually long lives." Those in her care had managed to live as long as forty-one months—remarkably long for a mouse of any ilk. Further, according to the study, her dwarfs showed delayed bone aging and reduced osteoarthritis, which she theorized were due to their growth-hormone deficiency.

The following year, John G. M. Shire, an endocrinologist at the University of Glasgow, took issue with the idea that dwarf mice age fast, noting in a brief report in *Nature* that such rodents in his lab looked perfectly normal at eight months of age. In 1976, Gary Schneider, a researcher at the University of Massachusetts at Worcester, chimed in, reporting that his Snell dwarfs didn't show any signs of the kind of immune deficiency that Fabris proposed had accelerated their aging.

Andrzej Bartke, then at the Worcester Foundation for Experimental Biology in Shrewsbury, Massachusetts, spoke up too. He had worked with the mutant mice since the early 1960s and was the world authority on the Ames dwarf, whose pituitary defect is nearly identical to the Snell's. He'd never seen signs of premature aging in the rodents, and said as much in a note published in a now-defunct periodical, the *Mouse Newsletter*, soon after Fabris's study appeared. "Nobody knew how long the dwarfs live," he told me in an interview. "But I knew they live longer than five months."

Then a more startling possibility emerged: Studies in rats suggested that growth-hormone deficiencies can actually retard aging. The counterintuitive data came from experiments in which researchers surgically removed young rats' pituitary glands in order to study the effects of eliminating key hormones. The procedure, called hypophysectomy, was very tricky—a scientist who was off by a quarter of a millimeter when drilling into a rat's brain would pierce a key artery and kill it. That made it difficult to collect enough data to draw strong conclusions. Still, W. Donner Denckla in the United States and Arthur Everitt in Australia independently carried out extensive hypophysectomy studies during the 1970s and early 1980s, yielding provocative signs of retarded aging, including delayed onset of tumors, kidney failure, and heart enlargement (a sign of cardiac weakening) in rats. The procedure also significantly extended life span.

It was only the second time that scientists had found a way to retard mammals' aging—the discovery of calorie restriction's life-extending effects in the 1930s was the first. Drilling into the brain, of course, wasn't very promising as an anti-aging therapy. But the hypophysectomy data raised the possibility that mammals' rate of aging could be slowed by altering hormone levels. The data hinted that dwarf mice, whose hormone deficiencies resemble those induced by hypophysectomy, may well be long-lived.

But that possibility still seemed far-out to most scientists during the 1980s. Besides, it was possible that by lowering growth-related hormones, hypophysectomy merely protected rats against hormonally driven illnesses, such as certain cancers, and thus extended their lives by delaying the onset of diseases rather than by actually putting the brakes on aging.

Still, two gerontologists at Maine's Jackson Laboratory, David Harrison and Kevin Flurkey, couldn't resist taking a look at the dwarfs' life spans in the mid-1980s. Harrison, a husky, big-boned man whom you might mistake for a North Woods lumberjack until you hear him enthusiastically explaining the subtleties of quantitative trait locus analysis, was fascinated by hypophysectomy's apparent anti-aging effects. But in the late 1980s, he and Flurkey, a postdoc in his lab, wound up throwing cold water on the idea that interfering with pituitary hormones could retard aging.

In studies with different kinds of male minimice, including the Snell dwarf and the "little" mouse, a growth-hormone-deficient rodent discovered at Jackson Lab in 1976, they found no sign of extended life span. (They didn't use females in the studies because studying females' hormonal changes during aging is formidably complex. As we'll see, their exclusive use of males led to a laboratory whodunit of great import.) Indeed, their data suggested that Fabris had been right after all about accelerated aging in dwarfs. You might even have said that the rumors of their resistance to death had been greatly exaggerated.

· · ·

One authority, however, still thought the dwarfs might be long-lived: Bartke. As an endocrine researcher, he didn't pay much attention to the gerontology literature. In fact, he simply didn't hear about the Jackson Lab scientists' dwarf study, which Harrison had reported in a chapter of a 1990 book on aging and genetics.

In 1984 Bartke had joined the faculty at Southern Illinois University in Carbondale, a Corn Belt college town that boasts the only geodesic dome R. Buckminster Fuller actually lived in—the famed inventor and popularizer of the domes was a professor at the university during the 1960s. Around 1990, Bartke began investigating the very opposite of dwarf mice: giant mice that had been genetically altered to overproduce growth hormone. Brought forth by genetic engineers in 1982, the giants were among the first "transgenics," or animals with implanted genes. They grow two to three times faster than normal mice on their way to becoming the hulking linebackers of the mouse world—they can be twice the size of their normal littermates.

Scientists quickly discovered that the giants tend to die young, typically before reaching their first birthdays. That didn't necessarily say anything about aging, though. The rodents' blood levels of growth hormone are sky-high, which scientists theorized had triggered life-shortening cancers and other diseases unrelated to aging.

But when Bartke, who had decades of hands-on experience with different kinds of mice, looked closely at the giants, he saw animals that seemed to be getting old before their time. His giants had come from an Ohio University expert on producing transgenic mice, Thomas Wagner, who supplied the rodents to Bartke for studies on how elevated growth-hormone levels affect fertility. (That was a hot topic in the Midwest, where dairy farmers inject cows with growth hormone to boost milk production, and it was suspected that the treatment had deleterious effects on reproduction.) At a few months of age, the giants look "like supermice, big and slick," Bartke said in an interview. "But after six to nine months, they begin to look old. They begin losing weight, getting gray and scruffy, and their backs are coming up"—elderly mice sometimes suffer from spinal curvature that gives them a humpbacked look. "I'd open a cage to take one out and immediately look at the label that showed its birthday" because it seemed unbelievably aged.

In 1993, Bartke and colleagues found signs of abnormally rapid brain aging in the giant mice. Meanwhile, German scientists reported that toward the end of the giants' short lives, their kidneys were shot, their livers were wrecked, and they often had multiple tumors. In light of such data, Bartke proposed that the giants really did undergo accelerated aging. But he got nowhere when he sought a grant from the National Institutes of Health to check it out. "I was an outsider who was new to the field" of gerontology, he said, "and I heard the same thing over and over"—that the giants' short life spans sprang from hormonally caused diseases, not accelerated aging.

He wasn't about to give up. A native of Poland who grew up behind the Iron Curtain, Bartke is a product of one of the world's most rigorous training grounds for challenging officialdom. Indeed, it's difficult to imagine how any bureaucracy could long withstand the pluck and persistence of a man who, as a gifted, young Eastern Bloc scientist during the 1960s, repeatedly wheedled permission, as he did, to visit the West from Soviet-style apparatchiks. Bartke wasn't particularly devoted to Poland's Communist regime either, and in 1967 he decided not to go home at the end of a training session at a U.S. research institute—he's lived in the States ever since and has long been a U.S. citizen.

A precise, soft-spoken, slightly formal man, Bartke is easy to spot at scientific meetings—he's the courtly, graying gentleman in the dark suit who, when meeting a woman, simultaneously startles and charms her by raising her hand to his lips. The Old World urbanity comes naturally—his father was a Krakow banker. But it tells you less than you might think. You probably wouldn't guess from Bartke's debonair manner that he has authored more than five hundred scientific papers—a towering mountain of work, including groundbreaking studies in both endocrinology and gerontology, that only an iron-willed workaholic could have produced. He's also a passionate outdoorsman with a flair for fishing contests. And he's not unlike that miracle of old-fashioned bioengineering, the mule. Bartke stubbornly plugged away for years doing important aging studies on a shoestring budget before his gerontology research received the attention it deserved.

He was born in 1939, the year Hitler's troops stormed into Poland to precipitate World War II. Bartke remembers little of the war. But he can readily bring to mind the neat rows of beetles he collected as a youngster and mounted in a glass case to proudly show his father, who had a passion for life science. In the late 1950s, Bartke was a standout in biology at Krakow's prestigious, six-hundred-year-old Jagiellonian University, and in 1960, when he was twenty-one, he was selected by the Polish Academy of Sciences to spend a year at a remote research station in the mountains of North Vietnam. It wasn't a great time to visit Southeast Asia—the Vietnam War was about to break out after Communist guerrillas backed by North Vietnam killed two American military advisers in July 1959. But Bartke was too busy filling specimen jars to pay much attention to the ominous developments. He even managed to bring back a previously unknown mite that now bears his name.

A few months after his return from Vietnam, one of his professors, a brilliant but notoriously brusque scientist, called Bartke into his office to tell the young man that he'd been nominated to attend the University of Kansas as a visiting graduate student—an extremely rare privilege. Bartke was instructed to think it over, consult his parents, and respond the following day—then he was abruptly dismissed. A few moments later the professor looked up to see the young biologist still standing there. Before the older man could voice his irritation, Bartke blurted that he didn't need to think it over or talk to his parents—the answer was unequivocally yes.

Among Bartke's first acquaintances at the Lawrence, Kansas, school was Robert Schaible, an amiable postdoc who would confer a great treasure on the visiting Pole. One day two years earlier, Schaible had glanced down at a litter of young mice in a lab at Iowa State University, where he was working on his Ph.D., and noticed something odd: Several of them were runts. He'd occasionally seen such undersized mice before, but never more than one in a litter. The multiplicity struck him as the possible work of a growth-stunting gene.

Schaible soon established that they did indeed carry a novel dwarfing gene whose effects are very similar to those of the Snell gene. He modestly decided against affixing his own name to the novel dwarf strain and instead dubbed it the Ames dwarf, because the mice had turned up in Ames, Iowa.

Like a modern Pied Piper, Schaible had a rabble of Ames dwarfs in tow when he moved to Kansas in 1962. Charmed by the tiny rodents, Bartke decided to focus his Ph.D. research on their endocrine deficiencies. When Schaible moved on to other things in 1963, he handed over his precious minimice to Bartke, who for many years after was the keeper of the world's only colony of Ames dwarfs. Bartke spent much of the next two decades detailing their hormonal and reproductive abnormalities.

After 1990, Bartke's work on the growth-hormone giants reawakened his long-simmering interest in the dwarfs' life spans. If he was right about the acceleration of aging in the giants, it followed that a dwarf mouse lacking growth hormone might age slowly.

Or did it?

A high-profile study published by the *New England Journal of Medicine* in 1990 had shown that giving growth-hormone injections to men over sixty boosted their muscle mass, increased their spinal bone density, and reversed the thinning of skin that occurs in old age. Partly funded by Eli Lilly and Company, a major seller of growth hormone, the report kicked off an anti-aging fad that, as I write this, is still going strong. Its leaders are physician-entrepreneurs who run clinics in places like Southern California that offer growth-hormone injections to aging clients, promising that the therapy will prevent everything from wrinkles to cancer. After the influential *New England Journal* had put its imprimatur on data suggesting that growth hormone has anti-aging power, Bartke's diametrically opposing idea seemed unsupportable, if not downright laughable.

Then one day in late 1993, two postdocs in Bartke's lab, Holly Brown-Borg and her husband, Kurt Borg, walked into Southern Illinois University's vivarium looking for old mice to use in a study on aging. "We were interested in two-year-olds," Brown-Borg, now at the University of North Dakota, recalled in an interview. "But we couldn't find any"—lab mice generally die before two years of age. However, "we saw there were still plenty of two-year-old Ames dwarfs around. So it was like, 'Hello?! Let's go upstairs and tell Bartke about this.'"

The lab chief was more than a little intrigued. Soon after, the three initiated a study in which thirty-four of the dwarfs were pitted against twenty-eight of their normal siblings in a slow race to the death. "Basically," said Brown-Borg, "we just sat there and waited for them to live out their lives" in hopes of finding a significant difference in life span between the groups.

The race took more than three years, but it got ever more exciting.

As the normal mice passed away, most of the dwarfs kept going—they wound up living a spectacular 50 percent longer. Two females even made it past four years of age, not far from a record for mouse longevity. The implications were historic: If the data held up under skeptics' fierce glare—and there were sure to be a number of fierce glarers—the Bartke lab would have extended the gerontogene revolution to mammals and simultaneously gouged a big hole in the accepted wisdom about growth hormone and aging.

But Bartke's heart sank shortly before he submitted the study for publication—he ran across the Jackson Lab study showing that dwarf mice are short-lived, making it seem as if his lab's result were simply a fluke. Bartke anxiously phoned the Maine lab to compare notes and quickly got through to Flurkey, who proceeded to recount a murder mystery that Bartke listened to with utter fascination.

• • •

Flurkey, a genial, owlish man, is one of the few gerontologists I know who was drawn to his calling by youthful intimations of mortality. Most gerontologists seem to start as young biologists who find aging intellectually engaging but not particularly relevant to them personally—it's difficult to feel the subject in your bones until osteoarthritis sets in. For Flurkey, the J. Alfred Prufrock feeling ("I grow old . . . I grow old . . . I shall wear the bottoms of my trousers rolled") hit home in a definitive way soon after he turned twenty. (T. S. Eliot, by the way, was twenty-seven when he penned *Prufrock*.)

"Every year through high school and college you're getting smarter, you're getting stronger," Flurkey told me. "Then it all stops. By your early twenties, you realize this isn't just going to keep getting better. In fact, if you look around you see that it's soon going to start getting worse every year. I hated that. And I thought, 'I don't know if I can do something about it, but I can try.'"

After training with several prominent gerontologists, Flurkey joined Harrison's lab in Maine to study aging in mice. In 1992, he initiated the lab's second study on Snell dwarfs' life spans—this time using both females and males. Like Bartke, he had a sneaking suspicion that earlier research showing they died young, including his own study with Harrison in the mid-1980s, hadn't told the whole story.

Indeed, over the years he'd become keenly aware of the tiny rodents' vulnerability to factors unrelated to aging that could shorten their lives and skew life-span studies with them. The depth of that concern came home to me when I visited the dwarf room at Jackson Lab. I found Flurkey's research subjects reposing on fluffy piles of fresh wood shavings while listening to a wide selection of calming, light classics played at low volume. Technicians in sterile booties and white coats periodically glided by to serve up fresh food and drink, taking care to speak in hushed tones. Since the dwarfs tend to get chilled easily, each was assigned to a normal-sized "caretaker" mouse with which it was caged to provide a big, warm friend to snuggle with.

Despite all the TLC, Flurkey's new study seemed to go awry. After about eighteen months, it was clear that his female dwarfs were astonishingly long-lived. But the males were dying at extraordinarily young ages.

Then one day he began to suspect foul play. After moving a dwarf and its caretaker from one cage to another, he heard squealing inside the new cage. "I looked over at the mice," he recalled, "and saw the dwarf on its back with the caretaker on top of it apparently going after its throat."

The fuss wasn't too surprising—both animals happened to be males, and male mice are known for squabbling, particularly after stressful events such as cage changes. But when pondering the episode later, he realized that it might have represented more than a minor dustup. A dominant male mouse, he knew, will occasionally kill pups sired by another male—estrus in females is suppressed by nursing, hence the infanticide enables the murderous male to bring the unfortunate pups' mother into heat so that she can conceive and bear his own offspring. Because dwarf mice look a lot like pups, could it be that they sometimes trigger this deadly behavior in their male caretakers?

Flurkey suddenly realized that this appalling scenario could explain his strange male-mortality data. He had caged his male and female dwarfs separately with same-sex caretakers—male dwarfs always had male companions. Correction: Puny, little male mice had always been caged with murder suspects three times their size. No wonder the females with their motherly caretakers far outlived the males. "I'd never seen an actual murder occur," Flurkey said. "But I'd often seen a male dwarf that looked fine one day turn up dead the next morning."

By the time Flurkey realized what was going on, only three of his study's male dwarfs were still alive. "But after we put them in with female caretakers," he said, "they went on to live almost as long as the female dwarfs." By the time Bartke phoned him, The Curious Case of the Dying Dwarfs had been solved.

Flurkey's complete report on the Snells' extraordinary longevity wasn't published until five years later—he had to conduct yet another long life-span study to obtain data unskewed by skullduggery. But the fact that both Ames and Snell dwarfs are long-lived was soon common knowledge in gerontology.

The exposure of the dwarfs' life-span secret marked a watershed. To be sure, gerontology's conceptual landscape had already changed with the discovery of worm gerontogenes, which showed that there are surprising exceptions to the rule that aging is mere genetic and biochemical anarchy. But skeptics could still reasonably argue that we higher, more complex animals are different.

Bartke and Flurkey left that argument in smithereens. What's more, the same month that the Illinois group's finding was published, scientists at the University of California at San Diego reported that they had isolated the Ames dwarf gene, called Prop-1. The San Diego researchers even pinpointed a single nucleotide—one of the three billion DNA "letters" that make up the genomic recipe for a mouse—that, when altered, endows mice with the dramatic life-span extension Bartke had demonstrated.

Bartke still faced a struggle to win over peers. After the landmark *Nature* report, the NIH awarded him fifty thousand dollars to further analyze Ames dwarf aging. But his subsequent request for a bigger grant was denied. Lingering doubts about his work's significance weren't surprising. The study in the *New England Journal* showing that growth hormone reverses signs of aging in elderly men seemed more credible and important than his contrary report about a few screwball mice.

But as time went on other kinds of stunted, mutant mice joined the gerontogene parade. One was a transgenic strain produced at Ohio University. Called the Laron mouse, the rodents carry disrupted growth-hormone-receptor genes, which effectively make their bodies deaf to the pituitary gland's hormonal commands to grow. (The name comes from Laron syndrome, a rare, inherited human disease that's caused by a similar genetic defect and was first described in 1966.) Laron mice show longevity gains of nearly 50 percent, on average, compared with normal siblings.

In 2001, Flurkey and colleagues reported that with proper care, the so-called little mouse, whose dwarfing is also caused by a defect in growth-hormone signaling, lives some 25 percent longer than normal mice do.

By around 2000, Bartke's once-unorthodox ideas about aging, hormones, and dwarfism were gaining credibility. One reason was that he and others had shown that dwarf mice don't just live longer, they also age with amazing grace. The memory powers of Laron and Ames dwarfs, for example, don't decline with age as they do in normal mice. And Brown-Borg, who had helped establish the Ames dwarf's longevity, showed that its free radical defenses are unusually robust.

Meanwhile, the growth-hormone-taking fad came under fire due to side effects such as high blood pressure, discombobulated glucose metabolism leading to diabetes, and carpal tunnel syndrome, which causes disabling hand pain and weakness. A 1999 study showed that giving growth hormone to critically ill patients doubled their death rate. Other data suggested that chronically administering the hormone increases the risk of cancer.

Bartke's cause also got a major boost from one of gerontology's most influential, big-picture thinkers, Richard A. Miller. With a frizzy halo of white hair and a white beard, a masterful stage presence (in college he was an amateur thespian), and a penchant for deadpan wisecracks, Miller could readily moonlight as a stand-up comedian, or perhaps a singularly mordant department-store Santa. He's known for penning comic poems called double dactyls (a verse form that might be called the limerick with a Ph.D.), and has written the best obituary ever for a mouse. Excerpt: "IdG1-030, the world's oldest living mouse, died peacefully in his sleep on November 15, 2001, at the age of 1450 days, just 11 days shy of what would have been his 4th birthday. Born and raised in a small plastic cage in Ann Arbor, MI, [his] parents had romped, poor but free, in the barnyards of Moscow, Idaho. . . . His great age did not, until the last day or two of his final illness, impair his zest for chewing wood chips or for doing pull-ups near the food hopper at the top of his cage. . . . IdG1-030 is survived by 9 half-nieces and nephews, 19 half grand-nieces and grand-nephews and 28 great half grand-nieces and nephews, all of them currently employed in the research field."

A professor at the University of Michigan School of Medicine, Miller was one of the first gerontologists to collaborate with Bartke— they jointly investigated patterns of gene activity that underlie the Ames dwarf's durability. Most important, though, he became the Thomas Paine of the revolution in thinking about the plasticity of aging that gerontogene discoverers had initiated. A much-cited essay he published in 1999 is arguably the revolution's most readable and persuasive tract.

Miller opened his version of *Common Sense* with a whirlwind tour of revelatory surprises in aging research—from Austad's long-lived island opossums to Bartke's dwarfs to the anti-aging effect of calorie restriction—then went on to argue that these bolts from the blue probably sprang from the same source: a gene-encoded "pacemaker" that synchronizes the many processes that turn young animals into old ones. If such a synchronizer didn't exist, he observed, it would have been virtually impossible for evolution, say, to have nearly halved the Sapelo Island opossums' rate of aging over a few thousand generations. Imagine the alternative: Senescence-retarding mutations in Austad's isolated opossums would have had to occur at roughly the same time in genes that control the aging of muscles, of eye lenses, of immune responses, and of myriad other bodily systems—a fantastic series of coincidences akin to a person winning the lottery scores of times over the span of a few years.

Miller's arguments eloquently brought home that a hierarchically controlled anti-aging module is likely embedded in the genomes of everything from worms to mammals. Indeed, as he put it, gene variants that could "slow down the aging process may well be hidden in our fields, orchards, and kennels, and perhaps even among our colleagues." The revisionist movement he helped spark greatly increased the anti-aging quest's plausibility, which in turn helped pave the way for biotechs, most importantly Sirtris Pharmaceuticals, to win major backing from investors in recent years. (It should be noted, though, that Miller, very much the rumpled academic, didn't set out to help gerontology spin off biotechs—it just happened that his insights served that purpose.)

Miller also has played a leading role in elucidating the tie between small body size and extralong life, a phenomenon that's not limited to mice—it also can be seen in dogs, horses, and, according to some studies, humans. Before taking a closer look at this fascinating variant of less-is-more, let's dispense with a possible point of confusion: It applies only *within* a species. That is, small dogs like Chihuahuas typically live much longer than big ones like Great Danes. As we saw earlier, just the opposite applies *between* species—bigger ones, like elephants, tend to live longer than smaller ones, like mice.

Before the longevity of dwarf mice was recognized, dogs represented the best example of the less-is-more phenomenon. The largest canine breeds' adult weight is roughly one hundred times bigger than that of the smallest miniature dogs, indicating that breeders have done a very thorough job of synthetically evolving little dogs with gene variants that stymie growth and, although the breeders didn't realize it, also promote longevity. Gerontologists have been wondering about dogs for a long time. In 1994, a University of Washington group led by Norman Wolf thoroughly explored the dramatic correlation between dog breed size and life span. For example, they found that only about 5 percent of Irish wolfhounds live past ten, while 50 percent of Boston terriers do. Big dogs were found to be afflicted at relatively young ages with cardiovascular disease, cancer, and other killers—the diversity of maladies indicates that they truly age quickly.

Of course, there are exceptions. (We're talking biology, after all.) For some reason, one giant breed, the Great Pyrenees, occasionally reaches fifteen. But the less-is-more pattern has been observed in an impressively broad array of species. Miller likes to show audiences a slide of a miniature horse that's fifty-seven—about double the typical life span of larger horses. In 2001, U.S. and British scientists independently discovered that dwarf fruit flies, which carry mutations very much like the daf-2 disruptions that extend worms' life spans, live 50 to 85 percent longer than normal flies.

The human data are more confusing, though. Consider the fact that women tend to be smaller than men and also to outlive them. Does that stem mainly from growth-related hormones? Estrogen's heart-protective effect in females? Testosterone's promotion of risky behavior in males?

One expert on the issue, San Diego management consultant Thomas Samaras, has compiled data suggesting that relatively short, light people tend to outlive tall, heavy ones. Among other things, his research indicates that professional baseball players, various famous people, and U.S. military veterans of height five feet nine inches or less live nearly five years longer, on average, than those taller than that—irrespective of whether they are heavyset or lightly built. A number of studies by other researchers suggest that tall people are more prone to various cancers than short ones.

But there are also reasons to doubt that short people tend to live longer than tall ones. For instance, shorter men appear to suffer from more heart disease than tall men, possibly because the former have smaller arteries, increasing the risk of high blood pressure and blockage that causes heart attacks.

The data on human dwarfs are also inconclusive. As mentioned earlier, people with leprechaunism die young. On the other hand, people with growth-stunting Laron syndrome have been found to live at least into their seventies. Further, the picturesquely named "little people of Krk" (pronounced "kirk," it's an island in the Adriatic Sea off the Croatian coast), a group of dwarfs with an inherited stunting gene that came to light in 1925, have been known to live to ninety-one.

Interestingly, at least nine of the dwarfs who played Munchkins in The Wizard of Oz in 1939 were still alive as of 2005 (some even had an agent and apparently were available for hire), while all the actors who played major, normal-sized characters had died by the mid-1980s. Such anecdotal evidence may not mean much—most cases of human dwarfism are not due to pituitary deficiencies, and thus aren't likely to be associated with long life. Still, one of the nine, Meinhardt Raabe, who played the Munchkin coroner, is on record as a pituitary dwarf. And he was not only merely alive, but really, most sincerely alive in 2007 at age ninety-two, when he appeared at the unveiling of a Munchkin star on the Hollywood Walk of Fame.

• • •

If hunting for practical ways to slow aging before the 1990s was like searching for missing keys in the dark, the discovery of gerontogenes in mammals could be compared to switching on a streetlight near the place they were lost. Basically, the halo of light covered the metabolic pathways that are altered in dwarf mice by their messed-up pituitary glands. But precisely which of the pathways holds the key to their extended life spans? Unfortunately, the pituitary isn't known as the master gland for nothing—the vast tangle of genes and enzymes it affects offered a confusion of riches for anti-aging researchers.

But a couple of pituitary-related pathways stood out as good starting points: the two that had also been implicated in the control of aging by the discovery that daf-2 mutations extend worms' life span.

As we saw in chapter 4, C. elegans' daf-2 pathway has two hormone-pathway counterparts in mammals. One centers on IGF-1, a growth-boosting hormone. The other one's main constituent is insulin, the key blood-sugar regulator. IGF-1 is thought to be a key link between dwarfing and long life—pituitary defects that lower growth hormone also reduce levels of IGF-1, which in turn limits growth and, at the same time, apparently helps to retard aging. Similarly, animals with low growth hormone tend to have low insulin, along with abnormally long life spans.

So here's the big question: Is it possible to tamper with IGF-1 or insulin pathways in a way that activates anti-aging effects without unacceptable side effects? After all, the FDA tends to frown on drugs that stunt growth, which is what lowering IGF-1 might do, or that tend to make people get fat, which is a common side effect of drugs that check insulin levels.

There are some animal data suggesting that IGF-1-lowering drugs might slow aging. In 2003, a team led by Martin Holzenberger at Saint-Antoine Hospital in Paris reported that mice that lacked functional versions of one of their two IGF-1 receptor genes lived 26 percent longer than normal littermates, yet were only slightly downsized and showed no reproductive abnormalities. (The life extension was largely limited to females, though.) A more recent Holzenberger study showed that genetically engineered mice with low brain levels of IGF-1 signaling have longer average, but not maximum, life spans.

But other data suggest that interfering with IGF-1 by itself doesn't extend mouse life span, and that fiddling with the hormone would backfire. Importantly, IGF-1 appears to have quite different effects in different organs. While low blood levels of IGF-1 are tied to slow aging, suppressing levels of the hormone in the brain might well have deleterious effects on learning and memory, among other things, according to some studies. Thus, drugs designed to reduce IGF-1 levels don't seem very promising as anti-aging agents at this point.

Meddling with insulin appears less risky, and as we'll see, insulin-modulating drugs are among the anti-aging quest's most promising compounds in recent years. In a 2003 study, C. Ronald Kahn and colleagues at the Joslin Diabetes Center in Boston showed that selectively disabling insulin receptors in fat cells extends the life spans of mice by 18 percent without adverse effects. While that life-span boost may not seem terribly impressive, it comes with a nice fringe benefit: Kahn's "fat-specific insulin receptor knockout," or FIRKO, mice remain svelte while eating all they want.

Many of us are already mucking about with our insulin pathways in a way that probably activates parts of our anti-aging modules. This miraculous intervention is regular exercise, which can slow the usual, age-associated rise in blood levels of insulin and glucose while lowering the risk of many diseases of aging.

Sadly, it appears that exercise doesn't really slow aging. That's no reason to stay on your couch, of course, given the proven gains in "health span" afforded by working out. (Exercise apparently doesn't switch on enough anti-aging mechanisms to retard the aging process, though it clearly does help ward off diseases of aging.) But if you're interested in giving the Reaper an extralong run for his money, it's impossible not to feel a little disheartened by the fact that rodent studies have pretty much quashed hopes that exercise extends maximum life span, the sign of a true anti-aging effect.

. . .

You may have noticed that I've dodged a burning issue in this mouse-lionizing chapter: Does the anti-aging module switched on by gerontogenes in mammals actually exist in humans in a form that, if manipulated in a sufficiently clever way, would slow aging?

It does seem likely that we all possess something like the module that gives dwarf mice wondrous longevity. After all, to geneticists mice are essentially just small, easily frightened people with tails. In fact, our major metabolic pathways are so similar to theirs that substituting key mouse genes with their human counterparts often causes no discernible change in the rodents.

But crucial details of the module in *Homo sapiens* can be best illuminated by studying our own kind. The natural subjects for such studies are the special humans in whom the module, or key parts of it, seems to be activated—in other words, centenarians, the subject of the next chapter.

Before turning to them, however, let's doff our hats to one last dwarf mouse—he not only set the longevity record for his species but also managed to bring about Andrzej Bartke's long-overdue moment of public glory. Known to the world as Mouse 11C, he was a forgotten, remarkably tiny Laron dwarf that was lost in the crowd at Southern Illinois University's vivarium until the morning of Wednesday, January 8, 2003, when Bartke walked through and noticed that a mouse had died overnight in one of the cages.

It was 11C. Checking its birth date, he discovered that it had quietly reached the age of 4 years, 11 months, and 3 weeks—about a year more than the astoundingly long-lived normal mouse Miller had eulogized two years earlier. Bartke was surprised, but like most scientists he has little interest in phenomena involving samples of one, and he thought little of 11C until the media got wind of the mouse's world record a few days later—the fact that a dwarf mouse weighing no more than eight paper clips had set a record for longevity wound up making international headlines. "A former student of mine in Kazakhstan even sent me a clipping from a Russian newspaper with the headline 100-YEAR-OLD MOUSE DIES IN THE UNITED STATES," Bartke told me.

Obviously, you can't believe everything you read in the Kazakhstan papers. An accurate headline would have read, MOUSE EQUIVALENT TO 190-YEAR-OLD MAN DIES IN THE UNITED STATES.

6

THAT OLD MAGIC

TIME TRAVEL HAS always struck me as one of sci-fi's most far-fetched conceits. Still, it hasn't seemed all that far from my realm of experience while I've hurtled through middle age. As events flash past at Internet speed, I've come to think that the time-machine concept isn't so much futuristic as it is outmoded. I can get an H. G. Wellsian rush just by turning away from the blur of the present and thinking back.

Loren Reid, the centenarian grandfather of one of my college friends, is better acquainted with such rushes than anyone I know. He's a retired professor of speech at the University of Missouri at Columbia, and some years ago he wrote a charming memoir, titled *Hurry Home Wednesday*, about what it was like growing up in the early 1900s in a small American town. A few weeks before he entered his 104th year in mid-2008, I prevailed on him to take me for a jaunt in time. Though somewhat frail and hard of hearing, he was as lucid as ever, able to conjure up nearly century-old scenes in phenomenal detail. Setting out from his memorabilia-filled house on a verdant, meandering street near MU's campus, we flitted back through the decades to visit his hometown, Gilman City, Missouri, population 600, circa 1915. His wife, Gus (Augusta), came along, occasionally steering us into interesting side trips. Like Loren, she looked to be in her mid- to late seventies. In fact, the retired MU English professor had turned 101 two weeks earlier.

Loren's parents appear in many of my mental snapshots from our outing to the way-back. His father was publisher, editor, and main reporter of Gilman City's newspaper, the *Guide*. A man of pluck and wit, Dudley Reid sometimes referred to his weekly as "the great moral rejuvenator," an appellation that may have had something to do with the fact that he was a die-hard Democrat in a predominantly Republican community. Loren's mother, another indefatigable doer, helped put out the paper and ran its business side.

Naturally Loren grew up perpetually ink-stained. At six he began hand-setting type—the same skill, he observed, that was mastered in youth by Ben Franklin and Mark Twain. A natural-born tinkerer, he became the family paper's chief linotype operator and mechanic during his teens. Linotypes were persnickety typesetting machines resembling colossal typewriters—the simple life ended for Loren as soon as he laid hands on one at age twelve.

Electricity came to Gilman City when Loren was in fifth grade. At first the power was on only from dusk to ten P.M. (eleven on Saturdays). Ten minutes before the lights went out there would be a warning blink, which on school-party nights precipitated a mad scramble to get the girls home before the streets went dark. The town's first automobiles were mostly for Sunday driving in fair weather—horsepowered conveyances made more sense on roads that often were little more than paths of axle-deep mud. In winter Loren never got enough of the pure joy of being towed on his sled behind horse-drawn sleighs up and down Gilman City's streets. Then he would dash home for a steamy bowl of navy bean soup seasoned with salt pork.

The Currier and Ives images I took away from talking to Loren and reading his memoir were juxtaposed with less idyllic ones that made it plain why his generation's life expectancy at birth was about fifty. The town center well, equipped with a hand pump and tin cup for common use by merchants and their customers, was topped up by street runoff in the rainy season. Relatives who came to visit from far away inevitably suffered bouts of diarrhea until they got used to the local water's brew of bacteria, and the Reid family took it as a matter of course when workmen cleaning their well once fished out a dead rat and "other additives" that Loren, in the wisdom of his years, decided not to go into. Families struck by typhoid and other deadly infections were quarantined, with dreaded yellow signs placed on their front doors. Minor cuts that became infected could be fatal—the age of antibiotics was still decades away. A number of his schoolmates' faces were disfigured by smallpox scars.

The remedies of choice for almost any illness were calomel, a mercury-containing purgative now known as a dangerous neurotoxin, and castor oil, mainly used today in brake fluid, paint, and other industrial products. The latter was sometimes followed by a chaser of corn whiskey mixed with coffee grounds to kill the taste. A gash might be treated with a poultice of bread, milk, and soda on a rag; no need to run for the doc—he could do little to ward off lockjaw or gangrene anyway. When the 1918 flu pandemic hit, Loren's father walked through the family business each morning dribbling formaldehyde, now classed as a carcinogen, on a hot coal scoop. The resulting fumes could "rip one's eyeballs out," Loren recalled, and they seemed to keep the deadly germs at bay. Many people dropped by to absorb a daily dose.

Reviewing the gauntlet of health hazards people faced a century ago, I find it remarkable that there are any centenarians at all today, much less a vibrant couple like the Reids. At the time I visited, they had been an item for more than eighty years and were about to celebrate their seventy-eighth anniversary. While needing help with chores and shopping, they were managing at home and keeping up with a wide circle of friends and relatives. Loren even gave me his e-mail address. He seemed at home in the Net-speed world.

· · ·

For centuries, students of aging have sought out extraordinarily long-lived people in hopes of finding out their secrets. It's now clear, however, that the wondrously wizened oracles of yore were rarely, if ever, as old as they claimed to be. The most famous longevity illusionist was Thomas Parr, a poor English farmworker who was interred at Westminster Abbey in 1635 after somehow convincing everyone, from England's leading physician to King Charles I, that he was more than 150 years old. In fact, "Old Parr" was probably no older than eighty when he died.

Claims of extraordinary longevity are so common—and so perennially evocative of mass gullibility—that debunking them emerged as one of the minor branches of demography in the late nineteenth century. William Thoms, an English writer who coined the term folklore, launched the myth-busting movement in 1873 with a bitingly skeptical tome, *Human Longevity: Its Facts and Fictions.* After exploding numerous claims like Old Parr's, he concluded that "scarcely one per cent" of purported centenarians could offer reasonably sound evidence that they were over one hundred.

Old Parr got most of the last laughs, though. The celebrated charlatan still lies among the immortals at Westminster (not far from Darwin); his portrait, by Rubens, no less, hangs in the National Portrait Gallery in London; and his story has been told in everything from encyclopedias to TV documentaries. Meanwhile, the far more obscure Thoms has had frequent cause to spin in his relatively modest grave in London's Brompton Cemetery. In 1973, for instance, *National Geographic* ran a cover story about the marvelously high prevalence of centenarians in places such as Ecuador's Vilcabamba Valley—the article was later shown to feature exaggerated longevity claims.

But since the mid-1970s a number of rigorous studies on centenarians have been initiated, beginning with the Okinawa Centenarian Study in Japan. While some of their findings have conflicted, the studies have revealed provocative patterns from the current outer reaches of the human life span.

One surprise has been the rapid proliferation of potential subjects for such research. In recent years, the number of centenarians has been rising 7 percent or more annually across the developed world, compared with about 1 percent for the general population. According to the Census Bureau, 37,000 Americans were over one hundred in 1990, compared with 50,000 in 2000 and 84,000 in 2007. Demographers project that there may be more than a million U.S. centenarians by 2050. (The great majority will be women, by the way—female out-number male centenarians by at least three to one.)

This dramatic change partly reflects the reduction of mortality among people before midlife that was achieved before 1950 by improved sanitation, vaccines, and other measures against infectious killers. But it has largely stemmed from a phenomenon that's harder to understand: Death rates of old people in the developed world rapidly declined after the mid-1960s. Indeed, just as baby boomers began burning their draft cards, their grandparents started quietly trashing actuarial tables. The result: Between 1950 and 2002 in the developed world, the odds of surviving to 90 after reaching 80 doubled, to 25 percent for men and 38 percent for women, and the odds of surviving from 90 to 100 tripled, to 3.6 percent for men and 6 percent for women.

Medical advances, such as the use of blood-pressure drugs to ward off heart attacks, doubtless played a role in the unprecedented change, as did lifestyle factors, such as less smoking. Some research also suggests that reductions in childhood infections decades ago have cut disease risks in today's seniors, possibly by lessening chronic, low-level inflammation that's thought to underlie many diseases of aging. But a definitive explanation has yet to be offered. When testifying in 2003 before the U.S. Senate on the trend, demographer James Vaupel of Germany's Max Planck Institute for Demographic Research simply identified its drivers as "the prosperity created by market economies" and "innovation based on research."

In any case, the declining death rates among the oldest old are so pronounced that one British bookmaker recently announced that it would no longer take odds on people living to 100 after reaching 90; the target age for such wagers would henceforth be at least 105. A spokesman for the firm told the *Guardian* that under the old rules, the bets were "starting to cost us a fortune."

• • •

Some years ago British writer Ronald Blythe observed that "the ordinariness of living to be old" was still novel, historically speaking, and that we would have to learn to grow old as we once learned to grow up. Now growing-older lessons come at us every time we turn on the TV, thanks to vendors of erectile restoratives, wrinkle creams, and the like. But marketers' recognition that a lot of us are of a certain age doesn't necessarily mean our culture's entrenched view of life after sixty-five as a period of wretched decline has faded much. After all, the relentlessly effervescent models in the ads for rejuvenalia typically look to be no older than fortysomething—it seems advertisers want to put as much distance as possible between their products and the specter of elderly decay. The bad rap on old age goes way back, and it won't die easy.

Nearly two thousand years before Shakespeare depicted life's last scene in As You Like It as "second childishness and mere oblivion,/ Sans teeth, sans eyes, sans taste, sans everything," Aristotle portrayed old age as a time of bodily chilling and slack passions during which men turn into cynical, distrustful, small-minded, cowardly, querulous "slaves to the love of gain." Understandably, most ancient thinkers saw no point in trying to prolong life. Indeed, Aristotle was an early member of what historian Gerald Gruman has dubbed the "apologist" school of thought on aging, whose adherents believe that it is neither possible nor desirable to extend life. The Greek philosopher maintained that aging and death, like other things he identified as part of the natural order, including the inferiority of women to men and the natural status of some people as slaves, are all for the best. Pursuing this idea with his usual rigor, he duly admired the clever way that nature makes teeth fall out in old age, a time of life, he noted, when they're no longer needed due to the imminence of death.

Another early apologist, the Roman poet Lucretius, perfected the why-prolong-the-misery argument by observing that there are only a limited number of pleasures in life, and once you've experienced them it's better to die than to face another tedious round. The Roman statesman Cicero was a little less gloomy. At sixty-two, a ripe old age during the first century B.C., he advised moderate diet and exercise to assure health in later life, and he maintained that older people could still cultivate their minds and characters despite their physical decline. He even welcomed the waning of sensual appetites as abetting virtue. But as Gruman drily commented, Cicero's argument that old age really isn't all that bad seems "rather defensive, like that of a lawyer saddled with a guilty client." Cicero wasn't particularly opposed to death, which he viewed as setting one free from a bodily prison on an imperfect earth. He reportedly faced death with great equanimity after being hunted down by soldiers of Mark Antony and his allies, who ordered the silver-tongued senator's assassination.

One of the few approximations to an upbeat take on old age in ancient times was an early example of black humor by the Roman philosopher Seneca. Tongue in cheek, he portrayed himself as an aged gentleman who visited one of his outlying estates and flew into a rage about the decrepit state of the grounds. Then he learned to his private dismay that the problem was simply that everything was as old as he was. Running across a withered old man near a doorway, he demanded of the caretaker, "Where on earth did you get hold of him? What possessed you to steal a corpse?" The corpse suddenly piped up: "Don't you recognize me? I am Felicio—you used to give me puppets at the Saturnalia . . . I was your playmate when I was little." Taken aback, Seneca replied, "The man is absolutely mad. Now he has turned into a little boy and playmate of mine. It could be true, though—he is as toothless as a child."

When Francis Bacon turned his clinical eye on the aged in the seventeenth century, he saw piteous creatures suffering from weak and trembling sinews, dry and wrinkled skin, cold blood, dull senses, dry and salty bowels, slow-healing wounds, and melancholy humors. Two centuries later, Schopenhauer reflected that "as we advance in years, it becomes in a greater or less degree clear that all happiness is chimerical in its nature, and that pain alone is real."

Practical-minded Victorian thinkers viewed the aged as burdensome, worn-out economic units in an era of scientific management and industrial competition. In 1881, New York physician George Miller Beard wrote that the practice of elevating the aged to positions of power is a "barbarian folly," and that the "enormous stupidity and backwardness" of governments everywhere is due to their domination by aging people whose "brains have begun to degenerate." The following year, English novelist Anthony Trollope captured the spirit of the times in *The Fixed Period*, a satire that portrayed a twentieth-century nation whose young leaders pass a law calling for all citizens to be given a brief period of retirement at sixty-seven and then peacefully euthanized.

In 1905, prominent physician William Osler famously alluded to Trollope's novel in a speech at Johns Hopkins University Medical School that warned about the danger of intellectual stagnation at schools filled with aging professors. Observing that men are comparatively useless after forty, he added that chloroforming them at sixty might be of great benefit to society. Osler was speaking in jest, but it was telling that many of his peers took him seriously, causing a furor. Press reports linked some twenty suicides to the widely publicized speech, and one New York City workman wound up in Bellevue Hospital after telling his wife that Osler was following him "with intent to chloroform."

Still, purveyors of gloom about late life have faced competition from more sanguine thinkers since the Renaissance, when it dawned on humanity with unprecedented force that the power of reason could change things for the better. The revisionist leader on aging was a sixteenth-century Italian, Luigi Cornaro.

The son of an innkeeper in Padua, Cornaro aspired to be recognized as a member of Venice's patrician Cornaro family, which traced its roots all the way back to consuls of ancient Rome. The haughty Venetians never accepted him as one of their own. But his unlikely claim was actually true—it seems he was descended from a Venetian duke named Marco Cornaro, according to research by U.S. art historian Douglas Lewis.

The Paduan parvenu outflanked the Venetians, though: He arranged for his daughter to marry into their family, ensuring that his descendants would have the credentials he was denied. He further tweaked their noses by refusing to hand over the dowry he'd promised as part of the deal.

A gifted wheeler-dealer, Cornaro made his fortune through shrewd investments, then enjoyed the life of the true Renaissance man. He was active in politics, patronized the arts, designed buildings, and sponsored the building of dikes and other public works. He even wrote a comic play. But his magnum opus was a book on graceful aging, *Discourses on the Temperate Life*, which he began in his midseventies and finished shortly before his death in 1566. Cornaro was extraordinarily long-lived for his times, but it appears that he exaggerated his age—he probably died in his mideighties, although many sources put him close to one hundred at death.

In any case, his *Discourses* represented a major milestone in thought about aging. Cornaro confessed in the book that he conceived his prescription for health and longevity—eat small amounts of simple food with a little wine, keep regular hours, don't get too cold or hot, stay out of strong sun and wind, avoid "bad air," forgo excessive sex—after nearly killing himself with overindulgence in his late thirties. Some of his ideas were archaic; he believed, for instance, in Hippocrates' doctrine that health springs from well-balanced bodily "humors." But his spirited depiction of the pleasures of a healthy old age, as well as his advice for attaining one, remain unsurpassed for their infectious enthusiasm and good sense. He was gerontology's Leonardo da Vinci.

Cornaro attacked apologists from four sides. First, he assailed their stoic fatalism, arguing, "I cannot believe God deems it good that man, whom He so much loves, should be sickly, melancholy, and discontented" in old age instead of "healthy, cheerful, and contented"—surely He wants us to keep going as long as possible. An ebullient stylist, Cornaro seemed to leap out of his seat to refute religious apologists' view that piety demands fixating on the afterlife to the exclusion of seeking to live long and well on earth: "Live, live, that you may become better servants of God!" he exclaimed.

Second, the time after sixty-five can be "the most beautiful period of life" for those who take steps to remain in good health. As evidence, Cornaro enthused that in his eighties he could climb stairs and mount his horse with the "greatest ease and unconcern." People who knew him, he cheerfully declared, would confirm "how gay, pleasant and good-humored I am." And no wonder: He still took great pleasure in reading, writing, and conversing with scholars and artists; traveling; strolling in his garden; overseeing the greening of waste lands; and, best of all, spending time with his eleven grandchildren.

Third, because attaining great learning and virtue is a labor of many years, growing old gives one a chance to do good works made possible by the hard-won wisdom of experience. An example of such worthy work by an old master, he added with charming glee, was his *Discourses*, "which I have pleasure to think will be of service to others."

Cornaro's final argument in favor of the good, long life was his most original and provocative: Those who live long enough, he maintained, can die a "natural death" without "sharp pains," passing away peacefully "after the manner of a lamp which gradually fails." As we'll see, recent studies suggest that he was onto something.

In the eighteenth century, the rise of science and the concomitant belief in progress inspired attacks on apologism that went far beyond Cornaro's moderate revisionism. In 1780, Benjamin Franklin speculated that medical advances may make it possible to prevent or cure all diseases in the future, "not excepting even that of old age." Franklin was so eager to witness such wonders that he half-humorously suggested to a friend that he wished it were possible, by "being immersed in a cask of Madeira wine," to put himself in a state of suspended animation and wake up a century later.

In 1794, the Marquis de Condorcet, a French philosopher and mathematician, echoed Franklin's optimism in an influential book on human progress. While in hiding from the rabid leaders of the French Revolution, Condorcet rhetorically asked, "Would it be absurd" to suppose "that the day will come when death will be due only to extraordinary accidents or to the decay of the vital forces, and that ultimately, the average span between birth and decay will have no assignable value?" He was captured a few months later but managed to evade the guillotine, it is believed, by taking poison supplied by a friend.

Carried away by techno-optimism, progressivists like Condorcet—and the later dreamers, schemers, and quacks who added a lucrative, pseudoscientific veneer to the anti-apologist agenda—arguably were (and are) closer in spirit to pre-Enlightenment magic thinkers like Ponce de León than to Cornaro and his sage, grounded optimism. And headlong, futuristic visions of scientifically enabled immortality have repeatedly offered fat targets for apologists, such as Thomas Malthus. The dour English economist's seminal work, *An Essay on the Principle of Population*, published in 1798, was partly written as an attack on Condorcet's speculations about extending life. Malthus maintained that if life expectancy rose much, overpopulation would follow, causing starvation and disease that would inevitably restore the status quo. His famously bleak view helped inspire early Darwinians' faulty idea that pro-aging, death programs evolved to clear away spent, old animals and make room for young ones.

Today the war between the apologists and their opponents rages on, although they mainly fight nowadays about the desirability rather than the possibility of significantly extending life. Since the 1970s, modern Malthusians have been sounding alarms about the rise of "greedy geezers," who are presumably gobbling up an unfair share of the world's resources. The critics maintain that anti-aging drugs would dramatically worsen this problem, causing disastrous population growth and crushing health-care costs. We'll come back to these big topics in the last chapter. But before grappling with them, it makes sense to examine what's known about the quality of extralong lives. Fortunately, the data on this issue are pretty heartening.

• • •

A few years ago researchers conducting the ongoing New England Centenarian Study reported some remarkable statistics: 32 percent of male centenarians and 15 percent of females had reached age one hundred with no signs of ten major age-associated diseases, such as heart disease, diabetes, or osteoporosis. In another study, the team found that only about one in five centenarians had a history of prostate, breast, colon, or other nonskin cancers. As Thomas Perls, director of the study and an associate professor at the Boston University School of Medicine, has noted, hundred-year-olds who have escaped major diseases of aging often look twenty to thirty years younger than they are. For them, one hundred is the new seventy-five.

The remarkable pattern also applies to the elite of the elite: supercentenarians, or people over 110. They are very rare—fewer than one hundred living supercentenarians are known at this writing—but the limited data on them show that they typically don't need to enter assisted-living institutions until they reach 105, five to ten years later than most centenarians. The upshot, says Perls, is that "the older you get, the healthier you've been."

Some other data paint a less sunny picture. Although a surprisingly large proportion of centenarians remain cognitively intact into their nineties, nearly two-thirds of 34 who were tested by Perls's team had some degree of dementia. A Danish team reported that of 207 centenarians they surveyed, about three-quarters suffered from cardiovascular disease and half from dementia. Only one was free from any chronic condition or illness.

Still, it's clear that centenarians really aren't like you and me—unless one of us happens to be among the one in ten thousand people who make it to one hundred. Gus Reid brought this home to me when I asked about her working life and learned that she was still fuming about the fact that she and her husband had been steaming ahead at full speed when they were forced to retire from MU at age seventy. "We decided to *protest* but not to *contest* it," she told me, smiling at the play on pro and con despite her flare of annoyance at the school's obviously silly rules.

It's tempting to think that such uncanny spunk is centenarians' reward for doing everything right—not least because if it were, we should all be able to attain similar durability by emulating their lifestyles. The first wave of major studies on aging people lent some support to this just-deserts idea. The Baltimore Longitudinal Study of Aging, launched in 1958 by Nathan Shock, known as the father of U.S. gerontology, has been one of the most informative. Still tracking more than fourteen hundred men and women, it has produced a mountain of data on what normally happens to us as we age: memory decline, bone loss, dulling senses—the litany is now known with disconcerting exactitude.

But the Baltimore study has also shown that lots of people decline far more slowly than the norm. This eye-opening discovery paved the way for the "successful aging movement," which took off in the 1980s as data rolled in from research on preventable risk factors for diseases of aging. The movement holds that many of the deleterious changes associated with aging can be held back by exercise, good nutrition, not smoking, staying mentally and socially active—the tried-and-true tricks of successful aging that you've no doubt heard about countless times.*

The importance of minimizing risk factors was underscored by a 1996 study on Danish twins born between 1870 and 1900 that indicated genes account for only about 25 percent of the variance in human life span. And some of this limited number of genes may have nothing to do with aging anyway, since they may increase risk of death early in life. Thus, choosing your parents wisely seemed to matter less than ever after the study appeared, at least when it came to longevity. And it seemed apparent that centenarians' longevity secret had mainly to do with their lifestyles and other nongenetic factors, such as exposure to environmental toxins.

But when researchers looked into centenarians' lifestyles, it became apparent that many of them didn't give a hoot about risk factors. Nir Barzilai, director of the Institute for Aging Research at the Albert Einstein College of Medicine in the Bronx, says the centenarians he has surveyed all deny having been exercisers, and 30 percent have been overweight. Some centenarians have been heavy smokers. One of the New England study's subjects said that for years he'd eaten bacon and three eggs for breakfast every day. Others said their daily diet had always included red meat. Missouri's Loren Reid is known in his family for routinely eating an entire of box of Oreos at one sitting. A few years ago, I took another centenarian, Catherine McCaig, out to lunch and watched with puzzled concern over my chaste salad as she polished off a frighteningly large platter of deep-fried fish and chips. When I asked about her usual diet, she replied with wonderfully appalling gusto, "I've always eaten whatever came along, as long as it wasn't moving."

To be sure, centenarians aren't all feckless Evel Knievels when it

---

*My favorite recipe for successful aging was devised by famed African American pitcher Satchel Paige: "Avoid fried meats which angry up the blood. If your stomach antagonizes you, pacify it with cool thoughts. Keep the juices flowing by jangling around gently as you move. Go very lightly on the vices, such as carrying on in society, as the social ramble ain't restful. Avoid running at all times. Don't look back, something might be gaining on you."

comes to righteous living. While some are overweight, virtually none is obese and about 80 percent of the New England study's participants said their current weight had varied little throughout their adult lives. Those who have smoked typically kicked the habit early. Alcoholism is nonexistent among centenarians. They have a remarkably strong tendency to stay active and engaged with those around them.

The most striking commonality among centenarians is their good humor and resilience, which, among those I've known, is manifested as a quiet, composed joie de vivre that makes people want to spend time around them. At age one hundred, Catherine McCaig took a long-awaited trip to her native Ireland, and soon after arriving she slipped and broke her hip. "I had a wonderful time in the hospital in Dublin," she told me, laughing at how the trip turned out. "There were all these girls planning to become nurses who wanted to take care of me. They were marvelous." One of the new friends she made in the hospital was still corresponding with her three years after her return.

The startling thing about all this is that there simply is no typical centenarian lifestyle. And centenarians' most prominent commonality— unflagging resilience—might result largely from genes that underlie basic temperament. Could genes be more important in determining human longevity than researchers had long thought?

In the late 1990s, Perls and colleagues decided to investigate this controversial idea by collecting data on the longevity of centenarians' siblings. Scanning local newspapers one day not long after launching the study, they came across a photo of a 108-year-old man in Quincy, Massachusetts, blowing out the candles on his birthday cake. They were delighted to discover a near-supercentenarian only a subway ride away. But what really grabbed them was that his 103-year-old sister was standing next to him in the photo. The researchers were even more excited when they contacted the pair and learned about their family: The siblings had a living 97-year-old sister. Two other sisters, then deceased, had lived past 100. And seven of their cousins had reached 100. Longevity-promoting genes offered the most reasonable explanation for such an astoundingly long-lived set of close relatives.

Some gerontologists dismissed the case of the long-lived siblings as highly suspect—probably just another example of longevity hype. The naysayers also worried that publicizing the discovery would undercut the successful aging movement's promotion of healthy lifestyles—why work at keeping fit if your late-life fate is decided by genes? But Perls's team forged ahead. In 2002, they published a telltale analysis of more than four hundred families that showed that centenarians' male siblings were at least seventeen times more likely to reach one hundred than their typical male peers were, and that their female siblings were at least eight times as likely. Two years later, Barzilai and colleagues at Albert Einstein reported a similar result: Parents of centenarians in their study had been about seven times more likely than typical peers to live past ninety.

A new consensus on human genes and aging emerged. It holds that important longevity-promoting genes do exist in people, and their benefits are most apparent in the very aged. Indeed, the effects of such genes are likely becoming more manifest as life expectancy rises—a longevity-enhancing gene might never get a chance to reveal its presence, for example, in a chain-smoking couch potato who works in a factory filled with airborne carcinogens.

Importantly, the take-home message here doesn't conflict with the successful aging movement. On the contrary, for the great majority of us, doing the right thing seems just as crucial as ever, because healthful lifestyles can to some extent emulate the effects of centenarians' longevity-promoting genes. Perls believes many of us can make it well into our eighties in good health that way. But remaining vibrant well past ninety seems to require a lucky combination of gene variants. Men are less likely than women to be so lucky, but when they are, the result is usually very impressive—eating Oreos by the box with impunity is the least of it.

• • •

The effort to divine centenarians' secret is now centered on their special genes. The first one was identified even before it was clear that hunting them made sense. In 1994, scientists at the Center for the Study of Human Polymorphism in Paris discovered that the "apoE4" gene, which produces a variant of a cholesterol-carrying protein that had been previously identified as a risk factor for both heart disease and Alzheimer's disease, occurs about half as often in French centenarians as it does in typical twenty- to seventy-year-olds in France. Similar findings in other groups of centenarians established that apoE4 is a kind of anti-gerontogene—longevity is enhanced by its absence, while its presence increases the risk of frailty in old age. (Most people carry other variants of the gene, apoE2 and apoE3, that are seen as mild longevity enhancers.)

While correlated with longevity, apoE appears to be more of a disease-associated gene than one that exerts broad effects on normal aging. But possible examples of the latter soon emerged. Perls made a splash with one of the first.

The son of a German physicist who worked in the U.S. space program, Perls developed a passion for his calling at age sixteen while working as an orderly at a Denver-area nursing home during the mid-1970s. Many of the elderly people he met defied the stereotype of the institutionalized aged as physically and mentally incapacitated, he later recalled, inspiring him to devote his career to geriatric medicine and the fight against "ageism"—the term was coined in 1968 by Robert N. Butler, founding director of the National Institute on Aging, to emphasize parallels between society's ill treatment of the aged and racism. After completing his medical training, Perls launched the New England Centenarian Study at Boston's Beth Israel Deaconess Medical Center in 1995. Three years later prominent gene hunter Louis M. Kunkel, chief of genetics at Children's Hospital in Boston, and Annibale Puca, an Italian postdoc in Kunkel's lab, teamed with Perls to look for longevity-linked genes in the study's subjects.

The group quickly won a $150,000 grant from the Ellison Medical Foundation, which software tycoon Larry Ellison, Oracle Corporation's founder, set up in 1998 to fund research on aging. Within two years they had a hit—a stretch of DNA on chromosome 4 appeared to contain a longevity-promoting gene, or possibly more than one.

In mid-2000, the researchers decided to form a biotech company based on their work. Dubbed Centagenetix, its ultimate goal was to develop drugs that emulate the effects of longevity-enhancing genes.

Their logic was similar to that worked out by founders of earlier gerontotechs: Because centenarians are resistant to common killer diseases of aging, medicines that mimic their genes might prevent or retard the progression of the diseases. And while Centagenetix might benefit from the buzz about basic discoveries on aging, it would focus on traditional, disease-specific drugs.

Forming the company proved trickier than expected. One complication arose when Beth Israel, which held rights to cell samples from Perls's centenarian research, licensed them to a potential Centagenetix competitor—the deal had to be unwound before the new company could go forward. By May 2001, however, all systems were go, and the company was launched with a $5 million, first-round investment by MPM Capital, a Boston venture fund specializing in biotech.

Unfortunately, the discovery that fueled early excitement about Centagenetix—the finding of a longevity-associated region of chromosome 4—led down a rocky road. In 2003, Perls and colleagues reported that they had pinpointed a gene variant on chromosome 4 that's correlated with long life. The finding seemed quite promising: The gene, for "microsomal transfer protein," or MTP, was known to play a key role in cholesterol metabolism, so it made sense that it would be linked to longevity. But fingering MTP didn't give the start-up a clear competitive edge: Big drug companies were already pursuing MTP inhibitors as novel agents to lower LDL, the "bad cholesterol."

Perls's group also failed to find a significant correlation between the MTP variant and longevity in an analysis of DNA from subjects in a French centenarian study. Danish and German teams reported similarly discouraging results in 2005. By then, however, Centagenetix was no longer a stand-alone company. In early 2003, it merged with Elixir Pharmaceuticals, a biotech based in Cambridge, Massachusetts, cofounded in 1999 by Cynthia Kenyon, the prominent discoverer of worm gerontogenes. The merger promised synergy by enabling the combination of research on human and animal genes that slow aging.

The conflicting results on the MTP gene and longevity weren't unique. By 2008 similar doubts had been raised about at least half a dozen other reports linking genes to centenarians' longevity. One problem was that longevity-associated genes of people in different parts of the world have been shaped by disparate diets, health risks, and other factors. Thus, gene variants that abet longevity in one set of centenarians may not do so in another.

A control-group issue also bedeviled the research: Due to the fact that hundred-year-olds' shorter-lived peers aren't available for comparative DNA analyses, researchers have generally compared centenarians' genes to those of control groups composed of living people who are decades younger. But the genetic shiftiness of human populations—caused, for instance, by the successive waves of immigrants to the United States from different countries and ethnic groups—means that there may be glaring differences between the genes of the younger controls and the centenarians that have nothing to do with longevity. It's easy to mistake such differences for life-span-associated hits.

There's no great way around these stumbling blocks. But there are ways to lower the risk of getting tripped up, and Albert Einstein's Barzilai has led the way in putting them into practice. Born in Israel, Barzilai followed his father, the dean of an Israeli medical school, into medicine, serving as chief medic of the Israeli army before getting his M.D. in 1985. A jovial, animated man who invariably spices his technical talks with jokes, Barzilai went on to train in geriatrics, endocrinology, and molecular biology before becoming director of Albert Einstein's Longevity Genes Project.

Barzilai has focused on Ashkenazi Jews, who are more genetically homogeneous than most populations because of their long history of social isolation—anti-Semitic discrimination through the centuries has helped limit the diversity of their gene pool. He has enlisted both centenarians and their children—and, importantly, the children's Jewish spouses. That enables his team to compare the DNA of the children, who carry longevity-enhancing genes inherited from their centenarian mothers or fathers, with that of their spouses, who are genetically similar but aren't likely to carry such genes. The strategy significantly improves the odds of spotting genes that are truly correlated with life span.

One of the first striking patterns Barzilai's team noticed was that their elderly subjects typically had high levels of HDL, the good cho-lesterol. They also had unusually large particles of both HDL and LDL, the bad cholesterol. Particle size matters: HDL, or high-density lipo-protein, absorbs and carries cholesterol to the liver for disposal, and larger HDL particles are thought to be more effective at that task than smaller ones. Large particles of LDL are probably linked to less heart risk because smaller ones are more likely to find their way into artery walls. In light of their initial discoveries, the researchers scrutinized their centenarians' lipoprotein-related genes and soon found several that stood out as possible longevity promoters.

One, for a protein called CETP, is especially interesting because it appears to reduce the risk of death from an array of old-age diseases. In fact, Barzilai and other researchers have linked the CETP variant to lower risk of atherosclerosis, heart attacks, hypertension, diabetes, age-related cognitive decline, and Alzheimer's disease.

Barzilai and colleagues have also found intriguing links between human and animal gerontogenes. In 2008 they reported that some of their female centenarians carry a gene mutation akin to ones that reduce body size and increase longevity in mice and other mammals. The human mutation dials down the activity of the cellular recep-tor for insulin-like growth factor-1, or IGF-1, the growth-promoting hormone regulated by the pituitary gland. Interestingly, the offspring of long-lived women who carry the mutation were found to be about an inch shorter, on average, than their peers—probably because their IGF-1 receptors are partially deaf to growth commands issued by their pituitaries. Barzilai's IGF-1 discovery, which was presaged by similar IGF-1 findings among other sets of centenarians, suggests that the evolutionarily ancient anti-aging module that's activated by animal gerontogenes also exists in our species.

Further support for that exciting possibility recently emerged in research by various groups, including Barzilai's, that have tied variants of a gene called FOXO3a to extreme longevity. FOXO3a is a human counterpart of daf-16, the hublike gene in nematodes that switches on an array of anti-aging pathways when the animals enter slow-aging mode. Like daf-16, FOXO3a appears to activate stress-resistance genes that harden cells against free radical damage and other insults.

In late 2008 and 2009, a flurry of studies indicated that longevity-associated FOXO3a variants in people over ninety-five have sweeping beneficial effects that look like retarded aging. The lucky people who carry them have significantly less heart disease, strokes, and cancer than noncarriers. Elderly carriers also have less trouble walking and fewer cognitive losses with age. Interestingly, the beneficial effects are closely tied to signs of youthfulness in carriers' insulin pathway, which regulates blood sugar—as most of us age, our cells gradually become less responsive to insulin, which sometimes leads to diabetes.

It's not clear that such genes retard normal human aging—they may boost longevity by averting a particularly broad swathe of diseases. But when poring over studies on CETP, FOXO3a, and other longevity-linked genes, I found myself thinking about the centenarians I know. Loren Reid, whose outsized sweet tooth would seemingly have brought on diabetes in a normal person, might well carry a protective version of the FOXO3a gene. And a CETP variant might account for Catherine McCaig's ability to consume lots of fried, fatty foods for more than a century with no signs of heart disease.

The fact that scientists are beginning to unravel centenarians' elusive secret doesn't mean that we'll all soon be able to follow them into supersenior citizenry. But the first drugs based on their genes may arrive sooner than one might think. At least two pharmaceutical companies, Roche and Merck, are developing pills to improve cholesterol metabolism by inhibiting CETP activity. The companies aren't about to jump into anti-aging R & D, but their experimental drugs may mimic the effects of the gene variant that Barzilai's team has shown is strikingly common among many centenarians.

• • •

Staring darkly into their crystal balls, modern-day apologists warn that anti-aging medicine will likely usher in a kind of night of the living dead thronged by figures resembling Tithonus of Greek mythology. He was a Trojan beloved by Eos, the goddess of dawn. Eos persuaded Zeus to bestow immortality on him, but she forgot to ask that he also be granted eternal youth. Naturally, Tithonus turned into a withered, babbling old man. Then Eos shut him up in a room, where he lay forever after. In a version of the myth that seems a little less grim, she turned him into a grasshopper.

Apologists love to trot out the Tithonus myth because it gives the impression that they're invoking ancient wisdom as they seek to conflate the consequences of anti-aging research with our worst fears about aging—most of us dread a long, bleak bedridden denouement more than death. (Or at least we say we do in surveys.)

But let's get real: Immortality isn't on any mainstream gerontologist's agenda—it bears the same relationship to research on aging that faster-than-light spaceships do to aeronautical engineering. Still, the Tithonus myth serves as a useful reminder of unintended consequences. The risk that life-extending drugs might lengthen the period of disabling disease before death for many people—call it the Tithonus scenario—warrants careful consideration.

No one can deny that the life expectancy gains of the past century have brought about a dramatic increase in the prevalence of chronic diseases toward the end of life. Instead of dying relatively young from swift killers such as infections and massive heart attacks, we have been dying ever later from slower-acting ones such as congestive heart failure, cancer, and dementia. Living longer has also lengthened our periods of affliction with nonfatal diseases of aging, such as osteoarthritis and hearing loss. A few decades ago, these indisputable trends led many medical experts to conclude that the Tithonus scenario was unfolding without any help from anti-aging drugs. Then in 1980, James Fries, a professor (now emeritus) at Stanford University School of Medicine, cast doubt on that pessimistic view in a classic *New England Journal of Medicine* paper titled "Aging, Natural Death, and the Compression of Morbidity."

Fries began by theorizing that average life span would max out at about eighty-five—he cited the 1961 discovery that cells can divide only a certain number of times as a possible reason. However, he went on, healthier lifestyles, medical advances, and other factors were enabling ever more people to postpone the onset of chronic disease and disability as they aged. As a result, the typical period of late-life morbidity was being compressed between ever-later dates of disability onset and the wall of maximum life expectancy—health spans would increase even as life spans topped out. And at the end of a vigorous life, increasing numbers of people would reach a state of body-wide fragility that sets up rapid decline and death after a minor stress, such as an infection. Fries called this demise without prior protracted disease "natural death"—a term reminiscent of Cornaro's sixteenth-century depiction of how the very old peacefully pass away.

Today few gerontologists buy the idea that there's a life-expectancy wall set by biological constraints. And the idea that the "Hayflick limit" on cell division (named after its codiscoverer, Leonard Hayflick) lies at the heart of the aging has lost favor. But the compression of late-life impairment has turned out to be one of the most surprising and important health-related phenomena of the past fifty years.

Between 1982 and 2005, the fraction of Americans over sixty-five who were chronically disabled dropped by more than a fourth, from 26.5 percent to 19 percent, according to federal surveys. Even better, the annual rate of disability decline has accelerated—it went from 1.8 percent between 1994 and 1999 to 2.2 percent between 1999 and 2005. A group at Duke University led by Kenneth Manton, one of the trend's top trackers, has estimated that if this profound disability decline hadn't occurred, the nation's annual Medicare bill would be $73 billion, or 17 percent, larger than it now is. That figure seems arrestingly large even to federal budgeteers.

It should be noted that this encouraging pattern isn't universal. A review of disability rates at age sixty-five and over during the 1990s showed that three of the twelve OECD nations (Belgium, Japan, and Sweden) had increasing rates. But there were more countries in which the rates were dropping.

Trying to explain the falling disability rates in the United States has caused a statistically significant increase in head scratching among health wonks. Possible contributors include the rise of barrier-free architecture that makes it easier for seniors to get places and do things; technologies that ease daily tasks, such as microwaves and drive-up ATMs; knee and hip replacements to avert disabling arthritis; earlier diagnosis and better management of chronic ills; better drugs to delay the progression of heart disease, diabetes, and cancer;

and a general rise in older persons' level of education, which is closely tied to healthy lifestyles. Since penning his landmark paper, Fries has helped compile data on the reduction in disability linked to fitness: Nonsmokers who exercise and stay lean typically postpone the onset of disability by 7.8 to 12.8 years compared with peers with bad health habits. No one, though, has a really good overall handle on why disability is dropping.

But what about the Tithonus scenario—are the oldest old also experiencing compression of morbidity?

Brace yourselves, apologists: The data on this group are the most encouraging of all. The proportion of functional, nondisabled Americans over eighty-five rose by a third between 1982 and 2005, more than three times the percentage gain during the same period among all those over sixty-five. To be sure, about half of the oldest old need assistance with daily living. But since 1982 the proportion of institutionalized Americans over eighty-five has fallen from 27.2 percent to 15.6 percent, a decline of 43 percent.

As we saw earlier, a remarkable proportion of the very old somehow stave off becoming disabled. Boston's Perls has reported that 90 percent of his study's centenarians were functionally independent at an average age of ninety-two. That doesn't necessarily mean they were free of chronic illness. Various studies suggest that the prevalence of such illnesses may be rising even as disability drops. Technology that helps seniors cope may partly explain this pattern. I suspect that the amazing resilience of really long-lived people also helps many of them remain independent even when grappling with chronic illness. In any case, it's arguable that compression of disability is what really matters in the aged, and, at least in the United States, the news on this front has been amazingly good for more than two decades.

Catherine McCaig, who died at 105 in 2002, refused to become disabled even as she experienced increasing morbidity toward the end of her life. After breaking her hip in Ireland, she had to use a walker. But when I visited her afterward in her small apartment, she often served me cookies she had baked herself, then regaled me, as always, with stories of her life, such as the time she caught a glimpse of Queen Victoria when the aged monarch visited Ireland in 1900. Once she

entertained me by accompanying herself on her piano while sing-
ing "Jesus Wants Me for a Sunbeam," a hymn whose title she found
hilarious. Each week a friend gave her a ride to a local radio station,
where she read recipes over the air for people on tight budgets. She
was inspiringly chipper to the end. One night she was rushed to the
hospital with intestinal bleeding; she died a few days later. Just hours
before she was stricken, I later learned, she had been packing for a
trip to visit friends in Nova Scotia.

7

CRACKING THE
LIFE-SPAN BARRIER

IN EARLY 1934, Depression-weary Americans were beginning to see tendrils of hope poking out of the bleak landscape. President Roosevelt's New Deal was bringing the economy back from the dead. Galvanized by the sight of elderly women scrounging food from garbage, California physician Francis Townsend had launched a crusade for government-funded pensions that would soon spur the creation of Social Security. Things were even looking up for the long-suffering Washington Senators, who had made it to the World Series the previous fall.

But one of the new year's most promising developments passed almost unnoticed. According to a brief article in the January 13 issue of *Science News Letter*, a Cornell University nutrition researcher named Clive McCay was nearing the end of a four-year study that had shown that rats' life spans were greatly extended when they were put on near-starvation diets.

Like many groundbreaking discoveries, McCay's seemed more of an anomaly than an advance when it came out. A glorious new chapter in nutrition science had been opened not long before by the discovery of dietary deficiencies that cause ancient scourges such as rickets, pellagra, and beriberi. Just five years earlier two scientists had won a Nobel Prize for the discovery of vitamins. In the wake of such progress it seemed almost subversive to suggest that a bunch of rodent Oliver Twists, raised on such short rations that their growth was stunted, could live radically longer than well-fed peers. Acknowledging the conventional wisdom, McCay stated in his initial report that the startling discovery seemed "little short of heresy."

The finding won him a measure of renown, and he was promoted to full professor shortly after its publication. But over the next several decades it was all but forgotten outside the back halls of science, a laboratory curiosity that didn't actually spark much curiosity. The few researchers who did study calorie restriction, or CR, as it is now called, largely focused on its anticancer effects. Most scientists were reluctant to risk wasting a lot of time probing a phenomenon that seemed as baffling and intractable as aging itself.

Even anti-aging hucksters were flummoxed. The fact that there was no clinical evidence that CR would extend human life span wouldn't have posed an obstacle to quacks, of course. But how to extract money from people by pushing special diets that cause relentless hunger pangs? Despite CR's status as the only reliable way to retard aging in animals, it represented the worst possible raw ingredient for manufacturing elixirs, a cosmic joke on the snake-oil salesmen of the world.

Now that anti-aging research is a hot area, it seems almost bizarre that CR spent decades on biomedicine's back shelf. It is as if physicists had treated the discovery of nuclear fission in 1939 as little more than fodder for after-hours bull sessions over beer. The unvarnished truth is that McCay showed that the rate of aging is incredibly plastic, and that it's supremely simple to brake it in animals whose inner workings aren't all that different from ours. In my view, no biomedical discovery of the past century was more astonishing and significant.

In recent years, drugs that mimic CR's effects without the need for radical dieting have emerged as the leading prospect for extending life span—you've probably taken at least a couple of them. (The most exciting ones are spelled out in following chapters.) In 2005, the RAND Corporation, an influential think tank, named such drugs as one of ten key medical advances benefiting the elderly that may arrive within ten to twenty years.

So here's a prediction: McCay, whom few people outside the study of nutrition and aging have heard of, will someday be recognized as one of the twentieth century's most important discoverers. He wasn't a genius with a capital G. But he had the probing contrariness of the original thinker, as well as great drive, grit, and integrity. His signature discovery had a lot to do with the fact that he never shrank from rowing against the flow. He may have sounded sheepish when he said his results represented borderline heresy. But he was sure of his data, and what they said came through loud and clear: His skinny rats had made history with a capital H.

• • •

Born in 1898 in rural Indiana, Clive Maine McCay showed an early interest in nutrition when, as a boy, he learned about calories from a government pamphlet and afterward became known in his family for enthusiastically announcing the caloric content of whatever he found on the table at mealtimes. He was orphaned in his teens—his mother died of cancer when he was eleven, and his father was killed in a train accident when he was sixteen. The tragedies didn't alter his trajectory. Always a top student, he earned his way through college and graduate school, never forgetting his father's advice to broaden his experience along the way with a variety of jobs. One summer he worked in the wheat fields of the Midwest, following the harvest from Oklahoma all the way to the Dakotas. Unlike most people, commented one of his childhood friends, he "always did what he set out to do. For instance, we dreamed of climbing in the Rockies. McCay did it."

At a time when ever more Americans were flocking from farms to cities, McCay went back to the land. In 1933 he and his wife, Jeanette, who also became a distinguished nutritionist, bought a run-down, fifty-five-acre farm a few miles from Ithaca, New York, where he was a young professor at Cornell. They gave it a picturesque name, Green Barn Farm, grew much of their own food, chopped wood for heat, and for years hand-pumped their water. Their kitchen became a nutrition lab. McCay's students were often invited out to sample experiments such as chop suey mixed with soybean sprouts and doughnuts fortified with brewer's yeast. One month the only meat that the fact-finding McCays ate was liver.

Wiry, quiet, and intense, McCay was an exacting but inspiring teacher. Once he caught one of his teenage nephews hurling apples at sheep grazing in his farm's orchard. A few days later, he beckoned the boy to stand next to the barn, walked back twenty-five feet, and proceeded to bombard him for several minutes with apples, rarely missing. Then McCay walked away without a word. "I have never again thrown anything at an animal," the nephew told an interviewer years later.

At age forty-five, when he was at the height of his academic career, McCay enlisted in the U.S. Navy and served in the Pacific during World War II; he helped develop improved emergency rations for stranded sailors and advanced to the rank of commander before returning to Cornell in 1946. After the war, his contrary streak came to the fore as he took issue with America's growing passion for soft drinks, white bread, and other nutritionally dubious foods. Always a strong advocate of natural foods, he warned about the growing general "failure to realize that what one eats affects one's health." As usual, he was so retro that he was way ahead of his time.

McCay's interest in CR dated from the mid-1920s, when he was a postdoc in the Yale University lab of prominent biochemist Lafayette B. Mendel, who was known for landmark studies on vitamins and other dietary essentials. A decade earlier Mendel had discovered that female rats could reproduce abnormally late in life if their food intake was restricted; he and his longtime collaborator, Thomas B. Osborne, also reported tentative signs that the rats' life spans were lengthened. One day McCay asked his mentor why he hadn't pursued a longer, rigorous study that might have established that dietary restriction extends rodents' life span. The senior scientist reportedly told him, "You are young. You try it."

Soon after, McCay made a stab at it while investigating the protein needs of hatchery-grown brook trout. Their brief lives enabled him to conduct a longevity study in months rather than years, as do such studies with rodents. Putting the fish on a low-protein diet that retarded growth, he found, doubled their life spans.

Digging into what was known about food restriction, McCay discovered that he wasn't the first to speculate that it might extend life. Like many of today's CR enthusiasts, he was particularly struck by what Cornaro, the famously long-lived Renaissance man, had to say on the subject. A translation of Cornaro's sixteenth-century *Discourses* had been published in America in 1917, acquainting U.S. nutritionists with his abstemious diet. McCay mused in one of his research papers that Cornaro's meager diet, like the one the Cornell scientist had used to extend life span in rodents, would have "barely prevented starvation."

One of the first scientifically compelling hints of CR's benefits had been reported in 1909 by Italian immunologist Carlo Moreschi, who showed that food restriction slows or prevents growth of transplanted tumors in mice. A few years later, Peyton Rous, a prominent U.S. cancer researcher, published a similar rodent study. Then came Mendel's discovery that female rats whose growth is slowed by food restriction remain fertile longer than normally fed ones.

But other research indicated that food restriction is harmful. In 1920, for example, scientists at the University of Adelaide in Australia reported that fast-growing mice live longer than slow growers, suggesting that a rich diet that fosters robust growth also boosts longevity.

McCay had doubts about the Australians' result, and he launched his fateful study on semistarved rodents in hopes of setting the record straight. Ironically, he was inspired by an erroneous idea proposed in the 1700s by French naturalist Comte de Buffon. Buffon theorized that animals' life spans were proportional to the time required for them to grow to maturity. If that were true, McCay reasoned, retarding growth during early life with a near-starvation diet should extend life span—he believed that some vital substance essential for life is consumed by growth.

McCay was right about the life extension. But the theory that stunting early growth is necessary to extend life was later proved wrong. Researchers exploded it years later by showing that when adult rodents are put on CR, their life spans are extended, albeit to a lesser extent than seen in those put on CR soon after weaning.

Despite his mistaken theory, McCay got a lot right that other scientists hadn't in prior studies on restricted diets. He solved the tricky problem of severely limiting food intake without malnutrition by using low-calorie rat chow formulated to contain adequate protein, vitamins, and minerals. He also employed a clever strategy to keep animals healthy when close to starving: Whenever any members of the group of calorie-restricted animals appeared to be failing, all were given more food until they had gained some weight.

About a year into the study, McCay and his student assistant, Mary F. Crowell, nearly saw their meticulously planned experiment collapse when two of their food-restricted females died during a spell of extraordinarily hot weather during the summer of 1931. While the early fatalities probably didn't have a major effect on their final results, at the end of the four-year study they were disappointed to find no significant difference between the average life spans of females on CR and those on normal diets.

But the male data were as clear and startling as a sonic boom. They showed, in essence, that McCay had cracked the life-span barrier seemingly set by nature. One of the two food-restricted groups of male rats lived 85 percent longer, on average, than the male control group did. McCay noted that his longest-lived rat endured for 1,421 days, nearly four years, while another researcher known for his rats' longevity had reported that none of his animals' lives had exceeded 1,250 days. In the study's extravagantly understated conclusion, McCay observed that the potential life span of rats "is unknown and greater than we have believed."

Over the following decade, McCay conducted two more major CR studies with rats, showing, among other things, that food restriction preserves youthful kidney and lung function in aged rodents, dramatically reduces their cancer risks, and even keeps them looking smooth and silky at ages greater than their species' average life span—elderly rats typically look as if they're having bad fur days. He also documented some downsides of CR, such as bone thinning in rats and a greater susceptibility to certain infections in young dogs.

In the 1940s and after, McCay took up many other nutrition issues: Do vitamin supplements extend rats' life spans? (No.) Does drinking loads of coffee shorten rats' lives? (No.) Does consuming soft drinks promote tooth decay? (Definitely.) He also wrote an acclaimed book on dog nutrition.

McCay died of a heart attack in 1967 at age sixty-nine, not long after retiring to Florida. His wife later wrote that although he always ate healthful food, exercised, and stayed thin, he basically worked himself to death. A few years after his death, there was a flare of public interest in his achievements, leading to glowing articles in national publications such as the *New York Times Magazine*. The focus of attention, however, wasn't CR. It was a soy-containing, protein-rich bread that he and his wife had dreamed up during the 1940s, endearing them forever to natural-foods aficionados. It was called Cornell Bread, or sometimes "the do-good loaf." You can find the recipe in *The Joy of Cooking*.

• • •

In principle, calorie restriction could raise life expectancy to nearly 120 if it worked in people as it does in rodents. That's average life span, mind you—maximum life span would hypothetically top 150.

But it's unlikely these huge gains could be attained. There's no proof that people respond exactly like rodents to long-term CR, and as we'll see, there are reasons to doubt that they would. Besides, achieving such gains in rats and mice entails putting them on stringent CR right after weaning; initiating CR in adult animals yields smaller gains. Similarly imposing heavy-duty CR on children would probably be deemed unacceptable by most parents, not least because it would likely stunt their kids' growth. And what parents could stand to see their offspring going hungry, even if the kids by some miracle were willing to put up with calorie restriction? Drugs that mimic CR's effects wouldn't obviate the problem, because by definition they would have profound metabolic effects, and administering potent drugs of any ilk to healthy kids is generally considered untenable.

Still, adults who religiously practice CR (or in coming years take drugs known to induce its effects, called CR mimetics) might well attain unprecedented longevity. Extrapolating from studies in which CR was initiated in adult rodents at various ages, one group of scientists estimated that if everyone reduced calorie intake by 30 percent beginning at around age thirty, life expectancy would climb by nearly seven years. Recall, that would be about twice the estimated increase in life expectancy that would follow the total elimination of cancer or heart disease.

Coincidentally, in 2006 a group of prominent gerontologists, including Robert Butler, founding director of the National Institute on Aging, stated that researchers should be able to develop an anti-aging pill that would delay the onset of all age-related diseases by about seven years. That "would produce the equivalent of simultaneous major breakthroughs against every single fatal and non-fatal disease and disorder associated with growing older," they wrote. "*And we believe it can be achieved for generations now alive.*" (Emphasis added by your aging author.)

Let's do a fast flyover of CR's remarkable effects to get a better idea of what the anti-aging drugs we're likely to see in the not-distant future might do. Keep in mind that the landscape below is lumpier than it looks when zipping across at high altitude. Zoom down and you'll see that CR's effects vary among different species, as well as among different strains within species. Mediterranean fruit flies, for example, don't live longer when put on CR. The life span of a strain of mice called DBA/2 is shortened by CR, probably because of a genetic quirk that increases infection risks.

There's no ambiguity, however, about the main thing: A strikingly broad array of organisms, even one-celled creatures like protozoa, live much longer when they consume considerably fewer calories than they typically do but still get adequate nutrition.* The list of CR responders includes yeast, water fleas, spiders, nematodes, fruit flies, guppies, hamsters, dogs, and probably monkeys.

Rats and mice have been the overwhelming favorites in the study of CR, whose discovery was presaged by the reports in the early 1900s that fasted lab rodents are strangely resistant to cancer. CR's antican-

---

*Various studies have indicated that CR's anti-aging effects in rodents stem from lowered intake of calories rather than of other dietary components. But there's evidence that reduced intake of methionine, an amino acid found in meat, fish, nuts, and many other foods, also extends rodents' life spans. While methionine-restricted diets may offer significant health benefits to people, little is known about their pros and cons, and they are difficult for humans to undertake because methionine is found in so many foods.

cer effects in rodents have long been one of the phenomenon's main attractions to medical researchers. In one study using four different strains of mice, 51 percent of elderly female rodents on normal diets developed tumors, while only 13 percent of their calorie-restricted counterparts did. In mouse models of cancer, CR has been shown to suppress tumors of the breast, prostate, intestines, and brain, as well as blood-cell cancers, such as leukemia. McCay reported that none of his calorie-restricted rats developed tumors until after they were put on normal diets at fairly advanced ages.

Some strains of rats often die of cardiomyopathy, an inflammation-linked degeneration of heart muscles. In a representative rat study, 95 percent of two-year-old rats on all-you-can-eat "ad libitum" diets suffered from the condition, while only 15 percent of those that were fed 40 percent fewer calories did. The same study showed that CR nearly eliminated kidney disease, another major killer of old rats, and attenuated the age-related degeneration of their adrenal and thymus glands, as well as of their livers.

Research on members of the Calorie Restriction Society, a group of people who practice CR, has shown that eating about 30 percent fewer calories than are contained in typical Western diets dramatically reduces blood pressure, LDL cholesterol, and other risk factors for heart disease. The hearts of middle-aged CR practitioners have been found to be youthfully elastic, resembling those of nonpractitioners about sixteen years younger.

One of CR's most important benefits is its ability to knock down chronic, low-level inflammation that's thought to drive many diseases of aging and is also likely a major component of normal aging. Along the same lines, CR keeps back overzealous immune cells behind inflammatory autoimmune diseases. Yet at the same time it retards age-related immune-system weakening.

In a study of mice that were genetically altered to develop a rodent version of Alzheimer's disease, CR halved brain levels of beta-amyloid—proteinaceous gunk whose accumulation is thought to be a key culprit underlying the disorder's massive loss of neurons. CR also protects rodents against the kind of brain damage caused by Parkinson's disease and strokes.

Aged animals on CR look much younger than peers on ad lib diets. "The skin of a [rat on CR] appears like that of a young one," McCay marveled in one report. Older rodents on CR don't lose hearing acuity, as do peers on normal diets. In both rodents and monkeys, CR greatly retards the inexorable loss of muscle tissue with age that represents a major cause of frailty in the elderly.

Despite all these pluses, however, it's not a given that CR compresses morbidity toward the end of life. That may sound like a paradox. But it's possible that CR postpones diseases of aging in people without abbreviating the amount of time the ailments subtract from health span. Thus, it's not immediately obvious that CR mimetics would help avert the crushing medical burden portended by the aging of the population—they might only delay its full measure of pain.

Happily, though, the literature on CR contains hints that it can, and often does, compress morbidity in addition to extending life. That means there's a good chance that taking CR mimetics would help many of us keep going in pretty good shape until a generalized state of fragility is reached that leads to precipitous collapse after a minor perturbation. In other words, it would enable an extralong, vibrant life followed by the kind of speedy demise we all hope for—one that doesn't leave us staring blankly for years at nursing home walls, confront our families with painful decisions about how much heroic medicine to try toward the end, or rack up huge bills that darken our loved ones' lives.

I picture this situation akin to the elegantly simple game of Jenga, which everyone from preschoolers to great-grandparents can play competitively. Players begin the game by setting up a compact tower of small, stacked blocks of wood. Then they take turns delicately extracting its embedded blocks and stacking them on the top to build it higher. The goal is to keep the structure standing until it is so full of weak spots that it suddenly crashes down at the lightest touch—the last player to touch it before the collapse loses.*

CR often seems to play bio-Jenga. In fact, about a fourth of rodents

---

*A similar analogy gerontologists often trot out when discussing compressed morbidity is the "one-hoss shay," a fictitious carriage described in a poem by Oliver Wendell Holmes Sr. Thanks to the fact that each of its high-quality parts was equally strong, it suddenly "went to pieces all at once" after lasting exactly one hundred years.

on CR suddenly keel over without warning at the end of their extraordinarily long lives, and close examination of their tissues after death shows no signs of organ degeneration. The cause of death in such cases isn't clear. University of Southern California gerontologist Caleb Finch speculates that because blood-glucose levels in CR rodents tend to be abnormally low, old ones may sometimes die from cardiac arrest brought on by transient drops in blood sugar.

Richard Weindruch, a University of Wisconsin CR researcher, once described the mysterious deaths of such calorie-restricted rodents to me in a particularly telling way: "When you open them up [to inspect their organs], they look great," he said. "The only apparent problem is that they died."

Perhaps we shouldn't be surprised if many obits in the approaching age of CR mimetics go something like this: "Jack Spry passed away on Friday at age 105 after a brief nonillness. Survived by his loving wife and kids, he was always the life of the party—including the one he attended on Thursday evening. His only major problem in later life was that it ended."

. . .

Life-span studies with rodents have told us a lot about aging. They've also turned many gerontologists into grizzled experts on Murphy's Law. There are an appalling number of ways for rats and mice wiling away their lives in cages to die prematurely. Insidious infections can traipse through. Previously unrecognized, late-onset genetic diseases can crop up. Subtle noises or smells that humans can't register may freak out the animals, shortening their lives. As we saw earlier, aggressive male mice sometimes kill their cage-mates when no one is looking. Such risks are magnified in life-span studies of rodents on CR, which tend to be especially long, affording more chances for things to go wrong.

Temperature gyrations rank near the top of the list of possible disasters. Killer heat, remember, almost crashed McCay's seminal CR experiment. Fifty years later, scientists in St. Louis conducting another CR study with rats lost many of their animals to overheating when an air-conditioning system failed. But history didn't exactly repeat itself. The latter calamity turned out to be serendipitous, providing an important clue about what's inside CR's black box just when scientists were finally beginning to lift its lid—the prelude to development of CR mimetic drugs.

The St. Louis study was carried out at a Veterans Administration hospital by a group that included Arlan Richardson, then at Illinois State University, one of the first researchers to investigate CR's mechanism at the molecular level. Now director of San Antonio's Barshop Institute, Richardson was dismayed when he came to work one blazing summer day in the late 1980s and learned that the thermostat governing the air-conditioning system in the hospital where his rats were housed had malfunctioned. When the lab heated up, an alarm that was supposed to go off also failed. As the hours crept by, the temperature in the rat room rose to somewhere over 90°F. By the time one of Richardson's colleagues discovered the accident, 86 of 128 rats had fatally overheated. They were twenty months old—middle-aged.

But there was a surprise: Three-quarters of the calorie-restricted rats had survived, while only 16 percent of the ones on ad-lib diets had. Apparently CR bestowed some sort of heat shield on rats. The fact that CR makes animals thinner was probably a factor—thin rats can shed heat faster than fat ones. But that didn't fully account for the oddity, because a normally fed, unthin group that had formerly been on CR also survived better. Thinking it over, Richardson and colleagues realized that a much more interesting phenomenon had likely protected the CR rats.

Research dating from the 1960s had shown that a wide variety of cells exposed to elevated temperatures react by churning out "heatshock proteins," which protect cells against heat and other forms of stress. It stood to reason that the ability to block cellular damage with heat-shock proteins fades with age along with everything else that goes downhill. If so, CR might postpone the downward slide, helping to make calorie-restricted rats relatively impervious to high temperatures. In 1993, Richardson and colleagues confirmed that, as suspected, the heat-shock response in CR rats remains youthfully robust as they age.

Their report provided some of the first evidence that switching on what's now called the stress response is one of CR's main anti-aging tricks. Fortuitously, the study appeared at about the time that a number of other scientists were finding that extralong life is correlated with cellular resistance to insults such as heat, radiation, and toxic chemicals. A couple of years later, the University of Colorado's Tom Johnson and colleagues postulated that both gerontogenes and CR extend life span by bolstering the stress response.

Linking the stress response to CR has clarified one of the most uncanny things about calorie restriction: It basically averts or delays what ails you no matter who you are. If you're a mouse that's genetically prone to develop kidney-wrecking autoimmune disease at an early age, CR waves its magic wand over your kidneys and doubles your life span. If you're a rat that tends to die of cardiomyopathy, CR fortifies your heart.

It's tempting to simply attribute this versatile beneficence to the fact that CR slows the fundamental aging process, which presumably works about the same in all creatures and eventually brings on diseases of aging. But this explanation both clarifies little and explains away too much.

Here's a better explanation that factors in the stress response: The induction of CR's diverse benefits across species by a common stimulus (reduced calorie intake) shows that an evolutionarily conserved module is being activated. A key part of its triggering mechanism is a calorie-intake monitor. Its anti-aging outputs largely comprise different aspects of the stress response. As with the anti-aging module triggered by daf-2 and related gerontogenes, the outputs have been customized during the evolution of different species to protect against the somewhat different sources of damage underlying senescence in their specific body types and life histories. That gives the module its versatile, cure-what-ails-you power.

In 1998, veteran calorie-restriction researcher Edward Masoro, who is credited with building the University of Texas Health Science Center at San Antonio into a major center for aging research, strengthened the connection between the stress response and CR by pointing out that CR's effects look a lot like "hormesis"—mentioned earlier, that's the lingering resistance to harm from toxins and other insults induced by small doses of such baddies.

First described by a German pharmacologist in the 1880s, hormesis was an esteemed concept in toxicology before 1930. Some physicians even applied it therapeutically. It was based on findings that at first glance didn't make sense (except to believers in Nietzsche's dictum that hurtful things that don't kill us make us stronger): Germicides in tiny doses speed the growth of microbes that they kill at higher doses. A little arsenic stimulates yeast metabolism. Low doses of X-rays boost animals' resistance to infections. Today it's thought that such examples of hormesis represent sustained activation of cells' molecular protect-and-repair systems after exposure to stress.

But the idea was tarnished by its association with homeopathy, a dubious form of alternative medicine based on administering vanishingly small doses of drugs, and by the advent of radium-spiked elixirs that supposedly acted as energizers via hormesis. In 1932, a millionaire playboy named Eben Byers died at fifty-one after gulping down more than a thousand bottles of one such elixir over three years, causing a lethal amount of radium to build up in his bones and eat holes in them. The nationally publicized case inspired a crusty editor at the *Wall Street Journal* to come up with an unforgettable headline: THE RADIUM WATER WORKED FINE UNTIL HIS JAW CAME OFF.

Hormesis was further linked to the weird and wacky in the 1950s, when earlier studies that had shown that low-level radiation could promote cellular growth via hormesis were twisted into plots for horror movies such as *Them!* and *Godzilla*, whose monsters supposedly got gigantic after exposure to atomic bomb radiation. The rampaging protagonist in *Attack of the 50 Foot Woman* also apparently underwent radiation hormesis, as did a certain amount of her clothes, after encountering a large, radioactive alien. All of which goes to show once again that what doesn't kill you makes you a lot stronger, bigger, meaner, and, on occasion, more scantily clad.

Still, hormesis is real and important, and it nicely explains anomalies that matter to all of us: Low doses of penicillin can stimulate bacterial growth; exposure to dirt and germs early in life lowers risk of allergies and asthma; and vigorous exercise, which seemingly should cause oxidative damage in cells, apparently strengthens them against it.

Hormesis may also partly account for the health benefits of consuming lots of vegetables, which contain natural, mildly toxic compounds that are known to stimulate aspects of the stress response in animals that eat them. In fact, scientists have long thought that children generally hate many vegetables as an evolved response to avoid such chemicals, which typically have telltale bitter tastes. We adults, however, have learned to love a number of bitter poisons that plants produce in their futile attempts to convince marauding omnivores like us to leave them alone. A neurotoxin called caffeine is one of them—wake up and smell the hormesis brewing.

Masoro and another gerontologist, Suresh Rattan at the University of Aarhus in Denmark, have made a compelling case that CR acts as a form of chronic, mild stress that continually invokes hormesis. That activates cellular systems that protect and spruce up DNA and other molecules degraded by aging. The fact that calorie-restricted rodents have chronically elevated blood levels of a hormone called corticosterone, a major driver of the stress response, suggests that they're right.

Hormesis isn't all there is to CR's anti-aging effect. For instance, CR has been shown to lower the output of free radicals from cells' energy-generating reactions, which, as Masoro has noted, doesn't seem properly classed as a hormetic phenomenon. But an appealingly large number of things about CR are consistent with the idea that it draws much of its anti-aging power from hormesis.

For instance, hormesis fits neatly with CR's evolutionary logic, which U.S. and British scientists independently proposed in the late 1980s. Their main insight was that CR triggers a "starvation response" that evolved long ago to help organisms outlast famine. The response is thought to entail cutting back on growth and reproduction in order to devote more energy to mitigating cellular damage, enabling organisms to age slowly and stick around to pass on their genes to future generations when conditions improve. Evolution might have cobbled together the starvation response by wiring nutrient sensors to various hormetic pathways.

This theory is widely accepted in gerontology, but not everyone in the field buys it. One skeptic is Steven Austad, who notes that unpublished data on mice living in the wild show that they rarely live long enough to benefit from putting the brakes on aging in order to ride out hard times. In fact, winter cold, predators, infections, and other mortal risks eliminate close to 99 percent of them in less than a year. Because they're long gone before reproductive senescence normally sets in, there would be no reason for them to evolve a slow-aging starvation response.

So how does Austad explain CR's anti-aging effects?

His alternative theory highlights the fact that when starvation looms, hungry animals forage more widely than they usually do. Nematodes, for example, are known for crawling off petri plates and making getaways during calorie-restriction experiments. Calorie-restricted rats tend to get frantically active, as if trying to desperately forage—they've even been known to run themselves to death on exercise wheels. A similar phenomenon may explain why some 80 percent of people suffering from anorexia nervosa become hyperactive.

When engaged in such "hard foraging," as gerontologist Caleb Finch calls it, animals branch out and eat unfamiliar things, exposing themselves to toxins not present in their regular food. Indeed, it's likely that most plants available to eat during tough times are rich in toxic chemicals, otherwise they'd already have been consumed. Thus, animals that engage in hard foraging probably have evolved very robust mechanisms to fend off damage from poisons as part of their starvation response. In Austad's view, the main reason that CR so handily extends life span in rodents is that they're geared by evolution to harden themselves against toxins as hunger sets in. The slowed aging that results from this erection of hormetic shields is an inadvertent fringe benefit rather than something selected for by evolution.

These two theories aren't necessarily mutually exclusive—the different evolutionary forces they contemplate may both have helped foster a conserved anti-aging module triggered by CR. In any case, both have the same basic bottom line: The power to slow aging is closely tied to the kind of advantage that evolution would favor—helping species ride out prolonged stresses that have been coming at them for billions of years. Indeed, the ability to dial up durability seems so handy that it's probably been around as long as living things have. And whenever evolution has been called on to confer it on a new species, it has been able to recycle existing metabolic machinery.

All this explains why different types of extraordinarily long-lived animals bear striking resemblances. For instance, both normal mice on CR and long-lived dwarf mice have abnormally low blood levels of IGF-1, insulin, and glucose, as well as reproductive deficits, unusually small body size, slowed maturation, and low average body temperatures. Some of these commonalities, such as low insulin and body temperature, have been observed in other long-lived animals, such as naked mole-rats, as well as in long-lived humans.

Still, nothing is simple in biology. Different methods of restricting worms' food intake, oddly enough, have been found to activate somewhat different sets of genes and metabolic pathways in the animals. And CR has been shown to extend the already-long lives of Ames dwarf mice, which indicates that it switches on at least some anti-aging mechanisms in the rodents that are separate from those activated by their gerontogene. It's also notable that long-lived dwarf mice tend to get fat as they age, while calorie-restricted mice are very lean.

Evolution has plainly composed variations on its life-span-extension theme. But the commonalities stand out more than the diversity. The stress response comes closest to qualifying as a standard feature of longevity enhancement, and because of that, hormesis theory arguably has more explanatory power than any other big-picture idea about CR. If you tilt the theory slightly and squint at it sideways, you can even get it to explain why the giant movie monsters of the 1950s were terribly hard to kill: Besides undergoing radiation hormesis, they were so busy ravaging the landscape that they had no time to eat and so were probably calorie-restricted as well—they were doubly hardened by hormesis.

. . .

Some party-pooping gerontologists argue that calorie restriction, and by extension CR mimetics, is likely to confer only minor human life-span gains, or maybe none at all. They have their reasons.

First, CR's well-known downsides loom large to the pessimists: It can make you so skinny that strangers think you're sick, thin your bones and muscles, make you feel too tired to exercise, make you painfully sensitive to cold temperatures, make women infertile, and slow wound healing. It's contraindicated for pregnant women and growing children. It lowers white-blood-cell counts and has been shown to make rodents more likely to die from influenza infections. Avoiding malnutrition on CR is tricky and time-consuming. CR also complicates the social pleasure of eating with friends.

Then there's the hunger. CR enthusiasts claim that it doesn't bother them much, and that you can deal with it by eating lots of high-bulk, low-calorie foods such as green leafy vegetables. But when I once tried CR for a few days, I found that it was impossible to fool my internal calorie monitor. After a while I felt like Templeton, the voracious rat of *Charlotte's Web*—when told by an old sheep that he would live longer if he ate less, he countered, "Who wants to live forever? . . . I get untold satisfaction from the pleasures of the feast." At least I was never as far gone as Clive McCay's poor canines: When he put dogs on severe, calorie-restricted diets, they went psycho—unlike normal dogs, they would eat dog meat, driven to cannibalism by craving.

Pessimists say that truly effective CR mimetics might well induce many of the ill effects of CR's strenuous dieting, possibly including relentless hunger. I don't buy this, for there are many examples of drugs that confer gain without pain that once seemed unavoidably tied to their benefits. People prone to paralyzing stage fright, for example, formerly popped tranquilizers or knocked back a couple of whiskeys before going onstage. In those days, wooziness and emotional dulling seemed the inevitable costs of suppressing the racing heart, sweating palms, and other manifestations of the "fight-or-flight response" that makes stage fright so awful. Now stage performers take small doses of beta-blockers to suppress the fight-or-flight response without any loss of mental clarity or emotional force. Thus, I think it will be possible to develop what Austad has termed "selective CR mimetics," drugs that confer CR's benefits without its main downsides.

The pessimists also maintain that CR's dramatic effects in rodents have misled us about how it would work in humans. Gerontologist

Leonard Hayflick, for example, argues that typical lab rodents are overfed couch potatoes that die young. Thus, subjecting them to CR merely eliminates life-shortening "overnutrition" and boosts their life span to what it would have been in the wild if they were protected from its usual hazards. It follows that CR doesn't really push the envelope on rodent longevity and is very unlikely to do so in humans.

But while mice in the wild tend to be leaner than lab rodents, they're obviously quite fertile and thus probably not truly calorie-restricted. Indeed, Austad has found that ounce for ounce of body weight, wild mice actually consume about the same amount of calories that lab mice typically do. He and colleagues have also demonstrated that wild mice on CR (actually, their grand-offspring) show lower mortality and compelling signs of slowed aging late in life. (For reasons that aren't clear, though, it increases their mortality early in life.)

Here is perhaps the pessimists' best shot: The evolutionary pressures that shaped the starvation response in rodents have never applied to humans, and so there's no reason to think we would react to long-term CR the way they do. Unlike small animals with fast metabolisms, we big mammals have relatively large bodily reserves that can see us through tough times without the radical move of forgoing growth and reproduction in order to slow aging. Further, if food is scarce in one place, far-ranging creatures like us simply move on to where it's not. Thus, evolution has had no reason to maintain the starvation response in our kind. It follows that the ancient anti-aging module that underlies the response's effects in rodents and other animals has probably lost its power in large, mobile primates like us—just as we and our fruit-eating primate relatives have lost the ability to synthesize vitamin C. (Over many generations, inherited traits that aren't favored by natural selection tend to erode away via random genetic mutations.)

Optimists counter that CR has extended life span in just about every species in which it has been tried, indicating that the evolutionary pressures that gave rise to and refined it are remarkably ubiquitous across niches, species, and time. Thus, the argument that we're an exception smacks of special pleading—a leap of gloom.

The case for optimism has also been strengthened by reports from two ongoing studies of CR's long-term effects in rhesus monkeys, one initiated in 1987 by scientists at the National Institute on Aging, and the other in 1989 by Weindruch and colleagues at the University of Wisconsin. Definitive data on whether CR extends the primates' life spans aren't expected for years. But like rodents on CR, the monkeys are very lean and appear to be protected from the pro-aging, inflammatory effects of visceral fat; their cells have remained youthfully sensitive to insulin; they don't show the usual age-associated rate of muscle wasting; and there's evidence that free radical damage in their muscles has been slowed.

In mid-2009, Weindruch's group made a splash by reporting that CR has indeed slowed aging in their primates—while 37 percent of control monkeys on normal diets had died, only 13 percent of the ones on CR had. The claim of slowed aging was somewhat controversial, however, because some of the monkeys on CR had died from causes deemed unrelated to aging, such as from gastric bloat; unless such deaths were excluded from the analysis, the CR group's longer survival wasn't statistically significant. Critics have argued that such deaths might actually be related to CR, hence excluding them may exaggerate CR's survival benefits.

But other data from the study strongly suggest that the CR really does slow aging in primates, probably extending their life spans by 10 percent to 20 percent, and that much, if not all, of that extra time is spent in remarkably good health. Compared with controls, the CR monkeys have greater lean muscle mass, significantly less age-related brain atrophy, half as much cancer, half as much cardiovascular disease, and zero diabetes. Overall, the rate of age-related diseases in the CR animals has been about a third that of the controls. And seeing is believing: Pushing twenty-eight years of age, about the average life span of rhesus monkeys, the ones on CR look to be lean, sleek, vibrant-looking adults, while controls have the hunched, frail, droopy-skin look of elderly humans.

Correlates of slowed aging have also been observed in people who have subjected themselves to CR for years, as well as in volunteers for an ongoing federally funded project to study CR in controlled clinical trials. In 2006, one of the first reports from the latter, called CAL-ERIE, or Comprehensive Assessment of Long-term Effects of Reducing Intake of Energy, showed that human responses to CR include lower insulin levels, just as in rodents.

One unintended study of CR has also shed light on its long-term benefits in people. It began in 1991, when eight intrepid earthlings launched a two-year voyage inside Biosphere 2, a three-acre glass-enclosed ecosystem in the Arizona desert that was bankrolled by Texas billionaire Ed Bass. Among them was sixty-seven-year-old physician Roy Walford, one of gerontology's most colorful figures. Sporting a shaved head and rakish Fu Manchu mustache during his later years, Walford first won national fame in the 1940s when he and a friend figured out a way to pick winning numbers in roulette and quickly won more than forty thousand dollars in Reno and Las Vegas. They used the money to buy a yacht and cruise the Caribbean.

His later adventures included walking across India to check meditating swamis' temperatures (he wondered whether they could lower them at will), exploring Amsterdam's underground drug scene, and breaking a leg trying to pop a wheelie on his motorcycle while riding along Santa Monica Boulevard in Los Angeles. He was also a gifted scientist known for pioneering studies on immune-system aging and CR. With Wisconsin's Weindruch, one of his protégés, he wrote a magisterial tome on calorie restriction, as well as highly enthusiastic consumer-health books on CR that make many gerontologists wince.

The biosphereans had planned to be totally self-sufficient, growing all the food they needed within the self-enclosed structure in order to emulate an extraterrestrial living experience. When their harvests fell short, they found themselves acting as guinea pigs in a CR study in which their calorie intake was chopped by 20 to 25 percent. (Gerontologist Richard Miller points out that the biosphereans' calorie intake might not have fallen if they had been willing to eat cockroaches, which multiplied wildly inside their structure and devoured much of their food.)

Walford closely tracked how they fared on their low-calorie, nutritionally adequate rations and later coauthored a report concluding that the biosphereans' hormonal, blood-chemistry, and other physiological changes had looked a lot like those of rodents on CR. Moreover, they kept up a high level of physical and mental activity. But a photograph of Walford after he had been inside the Biosphere for fifteen months, which was included in the study he published after it was all over, showed him looking emaciated and prematurely aged. A buildup of toxic nitrous oxide and a reduced oxygen level inside the structure might have contributed to the ill effects.

He died in 2004 at age seventy-nine of amyotrophic lateral sclerosis, the neurodegenerative disease that killed Lou Gehrig, which is apparently one of the very few illnesses made worse by CR. The night before he died, his daughter told the *Los Angeles Times*, he was fiercely working away on a Biosphere documentary video, his enfeebled arms suspended by pulleys over a computer keyboard.

• • •

For all the evidence suggesting that calorie restriction can significantly increase our health spans and life expectancy, there is little data in the CR literature to support the idea that it really slows human aging. That's another paradoxical-sounding statement that's really not paradoxical at all. As explained in chapter 2, proving that an intervention has anti-aging power requires showing, minimally, that it extends not average life span (life expectancy) but maximum life span. And that, of course, isn't feasible with CR in people, given that it would entail a study that goes on for something like a century.

But there is one human data set that some gerontologists regard as showing that CR has extended maximum life span. Researchers have known since 1949 that the calorie content of traditional diets on Okinawa, a multi-island prefecture of Japan, has been extraordinarily low even by Japanese standards. Data collected in the 1970s showed that adult Okinawans' calorie intake was 83 percent of Japan's average at the time. Thus, it seems likely that older Okinawans have spent most of their lives on what's considered a mild CR regimen.

It's probably no coincidence that Okinawans are the longest-living group of people in the world. In 1995, the longest-living 1 percent of Okinawans survived, on average, 104.9 years, nearly 4 years more than the comparable figure in the United States, and in Japan as a whole as well.

Okinawans also appear to experience compressed morbidity. They have 80 percent less breast cancer and prostate cancer than North Americans do. They have little heart disease; one autopsy of an Okinawan centenarian showed that her coronary arteries were virtually free of fatty deposits that precipitate heart attacks. Elderly Okinawans have about 40 percent fewer hip fractures than their U.S. peers. Their prevalence of dementia between ages eighty-five and ninety is half that of Americans.

Of course, other factors, such as the emphasis on vegetables and fish in Okinawans' traditional diets (but not in their increasingly Westernized ones), lots of manual labor, and the possibility that they carry longevity-promoting genes, may have contributed to their long lives and health spans. But these days it wouldn't seem the least bit heretical to suggest that their low-calorie diets have been one of the most important factors.

8

THE GREAT FREE LUNCH

THE FIRST SIGNIFICANT attempt to mimic calorie restriction's effect with drugs began with a flash of insight on January 6, 1997. That day, National Institute on Aging researchers Donald Ingram, Mark Lane, and George Roth were attending a conference on the biology of aging at the Ventura Beach, California, Marriott. They'd long studied CR, and among other things had helped to spearhead the first major study of primates on CR. For more than a year they'd been discussing the possibility of developing CR mimetics. But they were stuck. Six decades after the publication of McCay's seminal paper on CR, its mechanism was still too obscure to suggest compounds that might work.

During the meeting's second talk of the morning, Boston's Gary Ruvkun jolted the audience by revealing that the daf-2 gerontogene previously discovered in worms looks a lot like mammalian genes involved in insulin signaling. It was a preview of his lab's landmark paper on the isolation of daf-2, which forged a decisive link between anti-aging genes in lower animals and human biology.

Musing about the revelation under the hotel's palms a few hours later, the NIA scientists experienced a classic lightbulb moment. Gerontologists had long known that calorie-restricted mammals have low blood levels of insulin. But it hadn't been clear whether that is tightly linked to slow aging, or instead is an effect of CR with little bearing on life span. Ruvkun's discovery showed that interfering with insulin signaling—which is essentially what daf-2 mutations do in worms— can have profound anti-aging effects. That suggested that CR's reduction of insulin levels is a kindred phenomenon, and that drugs that lower insulin might brake aging as CR does.

After the meeting, Lane, now at AstraZeneca's MedImmune unit, scanned the medical literature for insulin-cutting drugs. Certain medicines for diabetes looked promising. But more interesting was 2-deoxy-D-glucose, or 2DG, a compound used to stifle cells' glucose processing in studies on energy metabolism. Structurally similar to glucose, 2DG jams enzymes that metabolize sugar, preventing its use as fuel.

Not surprisingly, heavy doses of 2DG are toxic, bringing on wobbly muscle control and stupor. But small doses, the NIA scientists reasoned, should emulate CR's relatively mild reduction of energy intake. The knock-on effects should include lower insulin and, with luck, slowed aging.

One of 2DG's big attractions was that it seemed likely to activate all of the metabolic pathways that cutting calories does. In drug developers' argot, 2DG hits an upstream target—its mode of action is akin to toppling the first domino in a set of branching rows to bring down the entire set. Since 2DG blocks one of the very first steps in nutrient processing, it comes into play about as far upstream as you can get in the hierarchy of pathways that trigger CR's anti-aging effects.

The researchers' excitement grew as they pored over studies detailing 2DG's physiological effects. It lowers body temperature in mammals, thus replicating another CR hallmark besides the reduction of insulin. It hampers reproductive cycling in female hamsters in much the same way that CR does in rodents. The compound slows tumor growth in rats, suggesting that, like CR, it lowers the risk of cancer.

Roth even ran across a diabetes researcher who'd taken 2DG while participating in a 1960s study on glucose metabolism. He'd been given small doses of the drug for a short time with no ill effects, although he recalled that it made him feel hungry. The fact that people had taken 2DG with no serious side effects was particularly important. A

CR mimetic might initially be targeted at a particular disease in order to elicit commercial interest and win FDA approval. But over time it was likely to be taken by lots of healthy people as an anti-aging drug, which meant that it would have to be extremely safe.

The researchers soon pulled together a short-term rodent study aimed at determining a safe dose of 2DG that could induce CR-like metabolic changes. Three groups of rats were put on food containing different concentrations of 2DG. Two other groups served as controls—their food didn't contain the compound.

In a clever twist, the researchers fed one of the control groups the same amount of food that was consumed at will by 2DG-dosed rats. If the "pair-fed" controls showed CR-like effects, the scientists would know that they were due to CR itself rather than the drug. That, in turn, would indicate that 2DG had probably done little more than make the rats' food unpalatable, causing them to calorie-restrict themselves. If that were the case, the scientists would need to rethink the project. They weren't interested in developing a CR mimetic that worked by letting people have their cake and eat it—and then spit it out in disgust.

Their initial finding was disconcerting though not terribly surprising: Five weeks into the study, four rats on the highest dose of 2DG had died, apparently from cardiac toxicity caused by the drug. The researchers dealt with the problem by administering the perilously high dose only on alternate weeks. That made the animals stop keeling over.

Despite the toxicity problem, the six-month study's overall results were encouraging. The lowest dose of 2DG seemed to have little effect on metabolism, and the highest dose obviously posed heart risks. But the middle dose had a Goldilocks glow. While having little effect on appetite or weight, it significantly lowered rats' body temperatures during most of the study, as well as their insulin levels—signs that it was mimicking CR.

When the researchers offered a report on the study to several journals for publication, though, it was roundly rejected. The concept of CR mimesis "was just not ready to be hatched in the mainstream literature," Ingram told me. "Even when we presented the data at meetings, many of our colleagues found it difficult to comprehend our enthusiasm about the approach."

Eventually, they found a home for the report in the *Journal of Anti-Aging Medicine,* a new publication that had little credibility with the medical establishment—its contents don't appear in PubMed, the comprehensive biomedical database maintained by the U.S. National Library of Medicine. As a result, the study that represents the Sputnik of CR-mimetic research is rarely cited in the medical literature. (In 2004 the *Journal of Anti-Aging Medicine* was reformulated as *Rejuvenation Research,* whose contents have been ruled worthy of admittance to PubMed.)

For a while in the late 1990s, however, it seemed that the medical world's most curmudgeonly gatekeepers would soon be bowled over by 2DG. In 1999, for example, University of Kentucky researcher Mark Mattson, now chief of the NIA's Laboratory of Neurosciences, discovered that the compound, like CR, protects rodents from the kind of brain damage caused by strokes and Parkinson's disease. But to win over the skeptics, the NIA group had to show that 2DG actually extends life span as CR does.

What happened next explains why you've probably never heard of 2DG before now. In late 1999, the NIA researchers launched what they buoyantly referred to as "the millennium experiment," a life span study of male rats given small, continual doses of 2DG in their food beginning at six months of age. Two groups of the animals were put on doses that, according to their earlier study, would lower insulin and body temperature without ill effects. As before, the experiment included two control groups, including a pair-fed one.

All went well for several months. But then animals on 2DG began to die prematurely. In fact, "the higher of the two 2DG doses, which had been completely nontoxic in the initial experiment, began to kill rats like flies," Roth later wrote. Once again the animals were dying of heart failure apparently caused by the drug. As the study progressed, even the lower dose, which initially had seemed to be working, failed to extend life span.

So it goes in drug research—the vast majority of compounds that look good in preliminary tests wind up as also-rans. But it would be wrong to dismiss the 2DG project as a total flop. It demonstrated that a drug can induce two of CR's most telling effects in rodents while allowing them to eat at will and maintain normal weight. That wasn't bad for an initial foray into uncharted pharmaceutical waters.

The fact that 2DG showed promise also inspired the NIA scientists to try related compounds, and in 2009 Roth disclosed at a meeting on experimental biology that a 2DG-like compound—mannoheptulose, a sugar found in avocados—had mimicked a number of CR effects in mice. Specifically, it reportedly improved insulin sensitivity and glucose metabolism; induced muscles to burn fat; and, most striking of all, extended life span by about 30 percent compared with untreated rodents without affecting weight or appetite. At this writing, however, the mannoheptulose findings haven't appeared in a peer-reviewed journal.

Perhaps the 2DG project's main payoff was to expand minds—the unabashed push into anti-aging research by a group of respected NIA scientists helped make it safe for gerontologists to venture into an area that had long seemed the reputation-destroying dark side.

. . .

As the idea of emulating CR with drugs got traction in the early 2000s, several biotech start-ups were formed to pursue it, including LifeGen Technologies of Madison, Wisconsin, and BioMarker Pharmaceuticals of San Jose, California. The NIA scientists also jumped in, forming a company to investigate drugs akin to 2DG—initially dubbed GeroTech, it's now GeroScience.

This first wave of CR-mimetic companies have been low-profile affairs compared with Sirtris Pharmaceuticals, the biotech juggernaut formed a few years later by Harvard's David Sinclair and venture capitalist Christoph Westphal to develop drugs based on resveratrol, the red wine compound with CR-like effects. They haven't been idle, though. GeroScience's team, for example, has explored pet foods containing possible CR mimetics, including mannoheptulose, with Procter & Gamble's pet foods unit. (I wouldn't be surprised to find out a few years from now that Fido and Muffy have quietly ushered in the age of truly effective anti-aging agents while their owners have been popping dubious nutraceuticals—food extracts that supposedly confer special health benefits.) The start-ups also deserve credit for beginning to transform the anti-aging quest from an endless guessing game into a fairly routine exercise in drug development.

That isn't to say that developing safe, reliable CR mimetics will be easy. Establishing such drugs' efficacy is likely to be the toughest challenge, and it's probable that the evidence they work will always be somewhat iffy. The case for efficacy will likely have three main elements: studies showing the drugs extend animals' life spans, clinical tests proving they induce CR-like metabolic changes in people, and longer-term trials confirming they reduce the risk of multiple diseases of aging and boost average life span among people who take them for extended periods. (We won't see definitive maximum life-span studies in humans that go on for a century.)

But this issue is minor compared with the challenges that defeated earlier attempts to slow aging. Importantly, scientists pursuing CR mimetics have effectively dispensed with the monumental problem of needing to understand the aging process. At first blush, that may sound like backsliding into willful ignorance—it contradicts the cherished idea in medicine that the best way to develop better treatments for ills is first to figure out exactly what causes them. The value of this strategy was famously underscored in the early twentieth century when scientists identified nutritional deficiencies behind diseases such as rickets and goiter, enabling outright cures. But when this eminently sensible approach was applied during the same era to the problem of slowing the aging process, it gave us goat testicle implants and radium-laced elixirs.

Such things are to be expected when you try to pull a rabbit from a barrel of snakes. George Williams, the evolutionary biologist who compared the anti-aging quest to the hunt for perpetual motion, was right about the aging process: It *is* a monstrous labyrinth. But he and other naysayers didn't foresee the maze buster offered by CR mimetics.

CR-mimetic developers don't need to solve the monster problem of how aging happens, because evolution has effectively solved it for them. It did so while fashioning CR's anti-aging machinery, whose elaborate downstream pathways are poised to carry out all the intricate metabolic adjustments necessary to extend life span. CR mimetics will take advantage of this preexisting mechanism, which still isn't well understood, to fend off aging and its myriad ills. In effect, the drugs represent the biggest free lunch in medical history. And given that compounds capable of emulating key effects of CR, including its ability to extend animal life span, have already come to light, it's arguable that the Great Free Lunch's appetizers are now on the table.

· · ·

Famed chemist Linus Pauling once said that "the way to get good ideas is to get lots of ideas, and throw the bad ones away." No one has taken this general principle more to heart than drug developers. They typically try to start out with a slew of promising compounds, knowing that's the only way to have a good chance of advancing one all the way to the medicine chest. CR-mimetic hunters face the same harsh statistics of attrition. Fortunately, biomarkers of calorie restriction—that is, metabolic changes closely tied to CR's life-span extension—enable researchers to identify lots of chemical starting points without doing a prohibitive multitude of lengthy life span studies.

As the NIA scientists realized at the outset of their 2DG project, an unusually low insulin level is probably one such biomarker. Low body temperature may be another. At this point no other CR biomarkers seem to be more telling indicators of slow aging than these two. As we saw earlier, the NIA group has found that extraordinarily long-lived men participating in the venerable Baltimore Longitudinal Study of Aging share two telltale traits with rodents on CR: They have unusually low insulin and body temperatures.

Other biomarkers have sprung from studies with gene chips. Sometimes called DNA arrays, the chips are credit-card-sized devices that read out the activity levels of thousands of genes in a sample of cells. (The word activity here essentially means the amount of protein manufactured from a gene's protein-making blueprint.) Gene chips enable researchers to track how the activity of various genes in, say, a mouse's muscles change as the animal ages, as well as to see the way that CR affects the changes. That means the chips can reveal global patterns of genetic activity correlated with CR's braking of aging. Researchers don't have to understand the mind-boggling dynamics that give rise to such patterns in order to use them like fingerprints in the quest for CR mimetics.

Chip-based biomarkers raise some knotty issues. For instance, increased gene activity during aging may be a sign that molecular damage is making things go haywire. Or it may mean just the opposite, that protective genes have been revved up to oppose damage.

Still, the chips have very intriguing things to say about aging. Stephen Spindler, a University of California at Riverside biochemist, has used the technology to show that CR can induce major gene-activity changes in mere weeks. His data also indicate that CR can repair preexisting damage to cells—it doesn't merely prevent further deterioration. Consequently, initiating CR (or beginning to take CR mimetics) late in life may do more good than has been generally thought.

Spindler has also discovered that a diabetes drug called metformin is remarkably good at mimicking CR's effects on gene activity in rodents. He and colleagues have reported that administering the drug to mice for eight weeks reproduces 92 percent of the changes that CR causes in the activity of liver-cell genes over the same period. In contrast, another diabetes drug called rosiglitazone (the brand name is Avandia), which has gross effects similar to those of metformin, reproduces only 13 percent of such changes.

Metformin's story goes all the way back to the Middle Ages, when nostrums made from a weedy plant called the French lilac were said to relieve the excessive urination accompanying diabetes. The plant, which is often called "goat's rue" (it's especially sickening to goats), and less commonly but more colorfully referred to as "holy hay" and "devil's shoestring," is processed today into alternative medicines to boost lactation. It's further claimed to increase breast size in nonlactating women, eliminate intestinal worms, and make children grow up strong. Satan also apparently swears by it for tying his Reeboks.

The plant's active ingredient of most interest, a compound called galegine, was isolated in the early 1900s, and molecular cousins of it, called biguanides, were introduced in the 1950s as drugs to help keep blood-sugar levels under control in diabetic patients. A popular biguanide, phenformin, was taken off the market in the early 1970s when it was found to increase the risk of lactic acidosis, a dangerous buildup of lactic acid in the blood. A few years later, however, a Russian group led by Vladimir Anisimov at the N. N. Petrov Research Institute of Oncology in St. Petersburg reported that chronic doses of phenformin significantly reduced tumor incidence in females of a certain strain of mice and—holy hay!—increased their average life span by 23 percent. The report got little attention, though—its data arguably meant little more than that a defunct drug can lower cancer mortality in one strain of tumor-prone mice.

Meanwhile, metformin, a cousin of phenformin that's less toxic than other biguanides, emerged as a mainstay of diabetes treatment. After two decades of use in Europe, it was introduced in the United States in 1995 under the brand name Glucophage by Bristol-Myers Squibb. A decade later, Spindler and colleagues spotted metformin's ability to mimic CR's gene-activity pattern. There's also some limited animal data suggesting that metformin can extend life span, and once again the source is a study by Anisimov's group in St. Petersburg that has received scant attention. The Russians' 2008 report stated that when hefty doses of metformin were added to the drinking water of a strain of mice (females only), their average life span rose by 37.8 percent and their maximum life span was increased by 10.3 percent.

Scientists still aren't exactly sure how metformin mimics CR effects, but it's clear that it affects some of the same insulin-related pathways that CR does. To understand how metformin might slow aging, you need to know a little about how insulin works—and how its failure to work properly is tied to aging.

When the blood's glucose levels rise after a meal, certain pancreatic cells release insulin into the bloodstream, delivering a signal to cells in muscle and other tissues to absorb sugar from the blood. As blood sugar then drops, a feedback loop shuts down the release of insulin. Unfortunately, this sugar-regulating system tends to get sloppy as we age, largely because our cells become less sensitive (or resistant) to insulin's message. This age-associated trend toward insulin resistance unfolds much faster in overweight couch potatoes than it does in lean exercisers.

To compensate for rising insulin resistance, the pancreas pumps out more insulin, and in advanced diabetes, the overworked pancreatic pump basically burns out, necessitating regular insulin injections. But even if we never get diabetes, our insulin levels typically rise as a function of how old and lardy we are. Age-related insulin resistance, and the accompanying rise in insulin levels, is closely tied to the insidious, low-grade, body-wide inflammation, touched on earlier, that goes hand in hand with many degenerative diseases of aging. As insulin resistance worsens, it leads to a kind of tipping point toward a disease called metabolic syndrome, a harbinger of many ills, including diabetes, heart disease, cancer, and Alzheimer's.

One of metformin's main effects is to make cells more sensitive to insulin's signal. That means that the pancreas doesn't need to pump out as much insulin to control blood sugar. Blood insulin levels then drop, and as a result insulin's other actions besides controlling blood sugar—it has diverse roles, including regulation of cell growth and protein synthesis—are also dialed back in ways that likely benefit health. In particular, reduced insulin signaling in fat cells is probably a key contributor to anti-aging effects observed in mice on CR and in long-lived mutant mice. (Recall that the FIRKO mice of chapter 5, in which insulin signaling is selectively disrupted in fat tissue, have extended life spans.)

Insulin's many functions still aren't completely understood, and scientists have long been puzzled by the fact that in worms, insulin receptor defects caused by daf-2 mutations extend life span, while in humans, similar glitches cause life-shortening ills such as leprechaunism. But it's clear that many diseases of aging in people are linked to insulin resistance and accompanying high insulin levels. That, plus the fact that attenuated insulin signaling often goes with long life, suggests that metformin's effect on insulin may brake a core driver of the aging process.

Some anti-aging enthusiasts are already blogging about the proper dose of metformin to add to their daily fistful of nutraceuticals. But it's not clear to me that it makes sense to chronically pop metformin in hopes that it would mimic CR. Among other side effects, the drug poses a slight risk of lactic acidosis. While very rare, about half of such cases are fatal, and the early symptoms, such as somnolence and malaise, are insidiously subtle and nonspecific, according to the medicine's package insert. (It should be noted that metformin's risk-benefit calculus is totally different for diabetes patients than it is for healthy people who take it as a possible anti-aging agent.)

• • •

Okay, it's time for another prediction: Gerontology will win its first Nobel Prize before 2020. My crystal ball doesn't have enough wattage to identify the particular advance that will win, but I'll hazard that it will involve the molecular biology of life-span extension. Several obvious contenders have appeared earlier in this book. One who hasn't until now is Leonard Guarente. For two decades his lab at MIT has probed how aging, gerontogenes, and CR work at the molecular level, turning out one arresting paper after another. Among other things, Guarente spearheaded research that led to the discovery that resveratrol, the celebrated red-wine ingredient, has CR-like effects in animals.

Guarente's career in gerontology began one day in January 1991 when two graduate students walked into his office at MIT and proposed joining his lab in order to look for genes that regulate aging. An expert on how genes are turned on and off, he was intrigued. But he had misgivings. The students, Brian Kennedy and Nicanor Austriaco, had pitched the idea after he'd challenged them to come up with a profound question to tackle while working on their doctorates. He'd urged them to think big partly because he was eager to do so himself. Guarente had spent most of his career doing filigree work on gene regulation, a pursuit that enabled him to rack up a solid track record, and he was worried about finding himself over the hill before he'd made a major discovery—or at least had tried to. But had he guided the youngsters to reach too high?

Hunting genes with major effects on aging clearly filled the bill as a momentous endeavor, and by 1991 there was reason to think such genes could be found—Tom Johnson had reported discovery of the age-1 gerontogene in worms three years earlier. On the other hand, Guarente was anxiously aware that such an iffy pursuit might leave his protégés empty-handed after years of work. So he struck a deal with them: If they hadn't made some progress after one year, they would switch to a more orthodox line of research to fulfill their Ph.D. requirements. He felt upbeat about his students' calculated risk until he ran across his department chairman in the hall and casually mentioned that his lab was launching a project on aging. The senior scientist blurted, "You're gonna what?!"

It took four years for Kennedy and Austriaco to publish a significant finding. But by then Guarente was thoroughly hooked, and, throwing caution to the wind, he'd converted his entire lab to the study of aging. By the end of the decade he'd emerged as one of gerontology's leading figures, and his lab had made national headlines by discovering what appeared to be a key component of CR's machinery. The discovery put him in the vanguard of the nascent quest for CR mimetics, and he went on to cofound Elixir Pharmaceuticals, a high-profile biotech company largely based on his research that for several years was the leading commercial effort to turn research on aging into anti-aging drugs.

The Guarente lab's research organism of choice, baker's yeast, didn't at first seem a very promising subject for studies on aging. While food scientists have long had a lively interest in the one-celled fungus, and geneticists are fond of prying into its DNA, it had never had much of a following in gerontology. *Saccharomyces cerevisiae* most recently shared a common ancestor with mammals more than a billion years ago. Its cells are so different from ours that they can thrive without oxygen. A stickler could argue that the microscopic fungus doesn't even age in the usual sense of the word, although "mother" yeast cells do get haggard-looking and bud off daughter cells more slowly over time, then cease budding before dying away.

A rapid talker with a shaved head and puckish sense of humor, Guarente seems at first blush to be one of the unreconstructed wunderkinder who whiz up and down the halls at MIT emanating thought bubbles filled with eigenvectors. Maybe it's his voice, a piercing nasal twang Fran Drescher might have had if she were male. Or perhaps that's just what happens to you when you've spent most of your adult life at MIT. In any case, it's a somewhat misleading impression. Lenny, as he prefers to be called, has a salt-of-the-earth side that sets him apart from standard-issue wonder kids.

The grandson of Italian immigrants, he grew up in Revere, Massachusetts, a downscale town just north of Boston known for its dog track and tacky beachfront. "I was precocious by local standards—I quit smoking in third grade," he recalls in *Ageless Quest*, a 2003 memoir of his early life and work.

His father kept food on the table with a clerical job at a local General Electric plant, but its low pay and prestige were a perennial disappointment to Lenny's mother, the family's dominant personality, who made no secret of her frustrated desire to move up the socioeconomic ladder. When Lenny's grade-school teachers told her he was gifted, she seized the chance to arrange his escape from the world of stickball-playing Marlboro men. After pushing for his admission to a prestigious private high school run by Jesuits, she got a waitress job to help pay for his tuition. Lenny pitched in by working at a local grocery store. Despite his part-time job and daily three-hour commute to and from school, he graduated as valedictorian of his class. His next stop was MIT, where he majored in biology; he went on to Harvard for graduate school and a postdoc.

Returning to MIT in 1981, he chugged up the tenure track by detailing how yeast genes work. But his marriage fell apart during the long haul, contributing to an early midlife crisis that, as such crises do when they work the way they're supposed to, summoned up the can-do, what-the-hell spirit of youth. It was an ideal state of mind for striking out into gerontology's great beyond.

The first major problem facing Guarente and his ambitious grad students was simple exhaustion. They'd set out to identify long-lived strains of yeast—ones that replicate more times than usual before passing away. Researchers had discovered in the 1950s that mother cells typically bud off about twenty daughters over a few days. That provided a way to measure the period during which a yeast cell passes from springy youth to withered decrepitude—called replicative life span, it's simply the number of times a mother cell buds off a daughter before dying. In principle, that made yeast a reasonable subject for aging studies. But yeast longevity studies had always required back-breaking bouts of sitting at the microscope. That's because you must count the number of daughters budding from a set of mothers growing on a petri plate and continually remove the newly formed cells with a fine needle, otherwise mothers and daughters get mixed up, making it impossible to establish the moms' replicative life spans.

For a while, Kennedy and Austriaco traded off doing consecutive twelve-hour shifts over the several days it took to complete one of the studies. But they were soon getting burned out. Then the lab's refrigerator came to the rescue: The scientists found that mother cells stop budding when put on ice and take up where they left off when restored to room temperature. That let them put aside the grueling experiments at will—no more all-nighters.

But another Everest loomed in the way: They realized that truly potent gerontogenes are probably quite rare, and that unless they got extremely lucky, it could take many years to find one using the searching technique they'd adopted. They were employing the same two-part strategy Klass and Johnson used to find worms' age-1 gerontogene. First step: Mutate your organism of choice with a DNA-clobbering chemical in hopes of engendering an abnormally long-lived mutant strain. Second step: If you're lucky enough to find a long-lived strain, zero in on it and try to identify its longevity-enhancing gene or genes. Step one alone seemed likely to require years of hunching over their microscopes, watching a zillion mother cells give birth.

Soon after getting started, however, they had a stroke of luck reminiscent of Alexander Fleming's. (Fleming was the penicillin discoverer whose breakthrough began with a neglected culture plate.) One day Austriaco took some stored yeast colonies from the fridge and found that their gelatinous agar medium had dried out. Starved and desiccated, many of the colonies had shriveled. But some survived and grew when given fresh media. What grabbed the scientists' attention, though, was that the survivors were strains that they knew from prior studies tended to live somewhat longer than normal.

That was really good news—it meant they could simply stick lots of mutated yeast in the refrigerator for a few weeks to cull out hardy, relatively long-lived strains. Then they needed to subject only that limited number of preselected strains to life-span studies in order to find truly long-lived ones.

The discovery of the nifty trick was doubly serendipitous, for it revealed a deep connection between the stress response, which was becoming a hot topic in gerontology during the mid-1990s, and genes that extend replicative life span in yeast. Just as gerontogenes in nematodes had been found to harden the tiny crawlers against all kinds of insults, the long-lived yeast emerging from the MIT lab's refrigerator turned out to be resistant not only to deathly dehydration but also to forms of stress such as heat shock.

By late 1993, they'd found a mutant strain with a single disrupted gene, as yet unidentified, that both increased life span by nearly 50 percent and caused sterility. (Yeast may not be very sexy, but they do sport a primitive version of sexual reproduction, enabling reproductively defective strains to be classed as sterile.) The scientists playfully dubbed the mystery gene UTH2, pronounced "youth too."

Next up, step two. Kennedy led the effort to isolate UTH2. Given the long-lived strain's sterility, he had a hunch that the underlying gene might be one of a number of sterility-causing mutations that had been discovered earlier. Sure enough, it turned out to be a variant of a previously identified sterility gene called SIR4. So much for their cute name—thereafter they addressed their gerontogene as SIR.

The letters stand for "silent information regulator," a designation based on the fact that SIR genes make proteins that deactivate, or silence, certain genes to prevent them from making unneeded proteins that might gum up a yeast cell's works. SIR4 was known to be part of a trio of genes—SIR2 and SIR3 are the other two—whose proteins combine forces to silence various genes.

Gerontogene in hand, the researchers next tried to figure out how it extends life span. They didn't get very far. But they did find a clue. It seemed that the longevity-enhancing mutation they'd induced didn't totally disable SIR4. Instead, it nicked the gene in a way that prevented the SIR4 protein from docking on genes it's supposed to silence. (Such docking is necessary for silencing.) That, in turn, caused SIR4 and its two SIR coworkers to wander off to DNA regions where they didn't usually go and silence genes there. The abnormally silenced genes, the researchers guessed, must include ones that normally make yeast moms get haggard after a few days of churning out daughters.

It was a nice hypothesis. But it raised three big questions: Which abnormally silenced genes are behind the moms' extralong life spans? What do those pro-aging genes normally do that causes mother cells to fade? Are there similar human genes that if quieted with drugs would slow aging?

Still, the team had clearly nailed a gerontogene. In early 1995 their discovery was published in *Cell*, a top journal, and soon after, a steady stream of aspiring, young gerontologists began flocking to the Guarente lab for research stints. One of the first was a driven, charismatic Australian postdoc named David Sinclair, who a decade later would become gerontology's all-purpose answer to Carl Sagan, Michael Jackson, and Bill Gates.

Sinclair met Guarente when the latter gave a talk at the University of New South Wales in Sydney soon after the *Cell* paper was published. Sinclair was about to polish off his Ph.D. at the university, and, like Kennedy and Austriaco four years earlier, he was casting about for an imminent supernova to hitch his wagon to. The idea of figuring out how aging works utterly enchanted him. Guarente was impressed with the eager young scientist and invited him to join his lab for a postdoc. But the lab's budget was tight, and so Sinclair was told he'd have to secure his own research funding.

Nothing daunted, Sinclair sold his beloved red Mazda Miata sports car to buy a plane ticket to Boston, where he wrangled an interview with a foundation that funds postdoc research by U.S. scientists—he gambled it would make an exception and support a gung-ho Aussie gone walkabout at the behest of an MIT professor. It did. Some days later, he breezed into Guarente's lab, greeting everyone with a cheery "Hello, mate."

A photograph taken soon after his arrival shows him with boyishly tousled hair and a bottle of beer in hand, smirking like a seventeen-year-old who has sneaked into a bar with a fake ID rigged up on his home computer. In keeping with the MIT lab's rollicking atmosphere at the time, which, according to Guarente, had earned it a reputation as the biology department's "frat house," one of Sinclair's first acts was to prominently display a news clipping from Australia that pictured zoo kangaroos mating while a bemused child and less-than-amused parents looked on. Some people didn't think it was all that funny, though, and it soon disappeared. But Sinclair proved himself a tireless, gifted investigator. "I piled a lot of things on his plate," Guarente told me in an interview. "I'd say, 'David, why don't you have a look at this.' He'd go off and work on it, and then I'd say, 'Let's have a look at that.' He never said uncle."

Sinclair soon became fixated on how the wandering SIR trio extends life span. After the *Cell* paper appeared, Kennedy had discovered that the trio gravitates to a region of the yeast genome that forms ribosomes, cellular components that manufacture proteins. That suggested an interesting possibility. Ribosomal DNA was known to be unstable in yeast—it tends to throw off gene pieces like a rattletrap losing nuts and bolts when taken for spin. Perhaps, the MIT scientists speculated, this ongoing disintegration is a root cause of yeast aging, and the abnormal silencing of key parts of ribosomal DNA by the SIR trio helps to prevent it. After all, nuts and bolts don't fall off quiescent rattletraps.

But exactly how did the decay of ribosomal DNA cause aging? The researchers needed to answer that question to have any hope of showing that their SIR work was relevant to human aging.

Then one day—October 28, 1996, to be exact—Sinclair flipped open his lab notebook and excitedly dashed off a six-page answer to the question under the hopeful heading "A theory on replicative senescence in yeast & other organisms." A year later he and Guarente submitted a paper full of data supporting the theory, which was so intriguing that it got Guarente on ABC's *Good Morning America*.

The theory holds that the unstable ribosomal DNA, or rDNA, tends to lose pieces over time that spontaneously form little circles. These circles inadvertently get replicated every time a mother cell dashes off a copy of her genome to make a new daughter. As a mother cell reproduces, so many of the useless little circles form that they eventually fatally overburden her genome-copying machinery. But in the lab's long-lived mutant strain, the wandering SIR trio's silencing of rDNA delays formation of the circles, slowing their toxic buildup.

The hypothesis made a splash partly because it resonated with a big idea that had been floating around for years: that genetic instability is a primary cause of aging in animals, including humans. After Sinclair reported how such instability makes yeast get old, it seemed that the Guarente lab's unorthodox bet on yeast as a divining rod for the root of aging was really beginning to pay off.

Two years later, in 1999, the lab directly tied SIR activity to animal aging when two of Guarente's most precocious graduate students, Matt Kaeberlein and Mitch McVey, singled out SIR2, the only member of the SIR-gene trio that happens to have counterparts in animals, as the one with the most anti-aging power in yeast. The finding focused intense curiosity on SIR2, and soon after, several prominent scientists were racing with Guarente's team to elucidate precisely how it silences genes to slow aging.

For a while it appeared that rivals had stolen a march on the MIT group. They reported that the SIR2 protein's function is to break apart a molecule called NAD (nicotinamide adenine dinucleotide) and then attach one of the resulting pieces to various other proteins. It seemed a plausible explanation of SIR2's modus operandi, since such molecular fragment transfers often convey signals inside cells.

But in late 1999, Guarente and Shin Imai, a Japanese postdoc who had arrived a few months earlier at the MIT lab, discovered an alternative SIR2 function that was far more exciting. In fact, the finding catapulted Guarente into the emerging high-stakes race to identify CR mimetics.

Imai had begun a study of SIR2 by assuming the rival scientists had got it right—that SIR2 cleaves NAD and sticks part of it to other proteins. Trying to add something to the picture, he'd hoped to identify which part of NAD gets broken off and transferred. With a high-tech scale called a mass spectrometer, which can weigh individual molecules, he measured the mass of protein molecules after SIR2 had added a NAD fragment to them. Since the proteins' original mass was known, simple subtraction told him the fragment's mass, which he'd assumed might reveal its identity.

Strangely, he found that the proteins had gotten lighter rather than heavier after SIR2 had supposedly augmented them. At first he and Guarente thought they'd screwed up—they were both novices at mass spectrometry. Then the penny dropped: The amount of mass each protein had lost was precisely the mass of a well-known molecule, called an acetyl group, that's often found attached to proteins in ways that alter their activity. That implied that SIR2's real function was to remove acetyl groups from proteins.

That fit nicely with earlier SIR discoveries, because it was known that detaching acetyl groups from certain proteins in cells' nuclei can deactivate genes. Thus, it seemed that Imai had shown precisely how SIR2 does its silencing thing. But over the next few days he and Guarente realized that their data had a more profound implication as well.

Their new excitement arose from the fact that they'd shown that SIR2 needs a helping hand from NAD to detach acetyl groups. But NAD, they knew, has another critical job in cells: It's involved in metabolizing glucose to release energy. This job is so important, the scientists guessed, that when NAD molecules are busy helping to generate energy, they largely neglect their SIR2 helper role. But when a cell doesn't have much glucose to process, the versatile NADs are free to assist SIR2 to silence genes. Thus, SIR2 largely sits idle unless a cell is low on glucose—which is exactly what happens during calorie restriction.

Putting it all together with speculative gusto, Guarente theorized that SIR2's real function is to act as sensor of a yeast cell's nutrient status, and, when its sugar fuel is scarce, to swing into action, removing acetyl groups in a way that silences aging-related genes and extends life span. In other words, his lab had found one of CR's master control molecules, and the fact that it has animal counterparts suggested that it might be a good target for drugs designed to mimic CR. Soon after Imai's study was published in *Nature* in early 2000, the chairman of MIT's biology department actually smiled at Guarente at a meeting and said, "Nice paper."

• • •

The rapidly unfolding SIR2 story seemed like a company waiting to happen after its CR connection became evident. Guarente had gotten some experience in biotech in the 1980s by cofounding BioTechnica International, a start-up that aimed to bioengineer improved food plants. Top-heavy with scientists lacking business experience, however, BioTechnica had capsized and most of its assets were eventually sold to a seed company. His next industry outing, however, brought him closer to the big time—in the mid-1990s, he'd served as a scientific adviser to Geron, the hottest gerontology spin-off of the decade.

Geron was formed in 1992 by Michael West, an heir to a family truck-leasing business who, after getting a Ph.D. in biology, had become a visionary entrepreneur possessed by a desire to rejuvenate aging tissues. In 1997, soon after Guarente had joined the company's advisory team, Geron's fame and stock price went ballistic when its scientists isolated the gene for telomerase, an enzyme ballyhooed as a kind of Zeus juice that could confer immortality on cells.

Telomerase's function is to prevent fraying of chromosomes, the coiled strands of DNA in cells' nuclei. It performs this vital role by preventing the deterioration of telomeres, protective structures at the ends of chromosomes that resemble the antifraying tips on the ends of shoelaces. Telomeres get progressively shorter as cells divide unless telomerase is on the job. But telomerase isn't normally active in most cells. As a result, their telomeres wear away over time, and at a certain point that causes the cells to stop dividing—they reach the "Hayflick limit," named after gerontologist Leonard Hayflick, codiscoverer of the fact that cells generally divide only a fixed number of times.

During the 1990s, telomerase enthusiasts speculated that telomere shortening, and the accompanying loss of cells' proliferative power, is the main driver of aging. Naturally, they hoped that telomerase-boosting drugs would retard aging. With the telomerase gene in hand, Geron seemed poised to bioengineer such breakthrough medicines.

Ironically, though, Guarente found evidence in yeast that telomere shortening actually does not drive the aging process, implying that the anti-aging buzz about Geron was off base. His association with the hot biotech didn't last long; it ended in 1998. But Geron's top managers didn't really disagree with him. At about the time Guarente left,

Geron deemphasized research on aging in order to focus on developing cancer drugs based on telomerase.

It was an eminently sensible move. For one thing, tumor cells abnormally activate telomerase to circumvent the Hayflick limit on growth, and thus medicines that block telomerase may halt their proliferation. (By the same token, drugs designed to slow aging by boosting telomerase may cause cancer.) Further, it was becoming evident from various studies—not just Guarente's—that while telomere shortening is involved in the senescence of various kinds of cells in the body, as well as in a number of age-related diseases, it isn't the chief driver of body-wide aging. For example, mice age and die long before their telomeres wear away.*

Guarente came away from his Geron experience with a renewed desire to start his own company as well as a lively sense of investors' growing interest in anti-aging research. In 1997, he ran across a kindred spirit with money—venture capitalist Cindy Bayley, a freewheeling biotech specialist who had also been thinking about starting a company on aging. She and Robert Nelsen, another VC at her firm, Chicago-based ARCH Venture Partners, advised Guarente that he'd have an easier time rounding up VCs to back a biotech on aging if he joined forces with another big-name gerontology researcher. It didn't take long for him to single out San Francisco's Cynthia Kenyon, the celebrated discoverer of the daf-2 gerontogene, as the ideal cofounder—she was the field's leading media star and Guarente had known her since her postgrad years at MIT. In 1999, the two teamed up to begin organizing a company.

Meanwhile, Guarente was continuing to enhance SIR2's luster as an anti-aging gene. In 2000, Su-Ju Lin, a postdoc in his lab, showed that CR failed to extend yeast cells' replicative life spans when their SIR2 genes were disabled—a sign that, just as Guarente had postulated,

---

*Telomeres' role in aging still isn't clear, though. In 2008 Spanish scientists reported that mice with artificially revved up telomerase showed signs of retarded aging. (The enhanced rodents also had to be implanted with cancer-blocking genes, because boosting telomerase normally causes tumors.) And in late 2009 a group at the Albert Einstein College of Medicine reported that some centenarians carry variants of the telomerase gene that appear to enhance longevity by keeping telomeres in good shape.

SIR2 lies at the heart of CR's machinery in yeast. Soon after, another Guarente protégé, Heidi Tissenbaum, demonstrated that implanting extra copies of a SIR2-like gene in nematodes extends their life spans. The finding showed that SIR2's anti-aging power exists in species that originated many millions of years apart in time (yeast and worms), and thus might even have been handed down to us—like all mammals, we carry a gene called SIRT1 that's quite similar to SIR2.

In late 2000, Guarente and Kenyon coauthored an ebullient review of advances in gerontology that essentially laid out the vision behind their nascent company. "The field of ageing research has been completely transformed," they declared in the high-profile paper, which appeared in *Nature*. "When single genes are changed, animals that should be old stay young. In humans, these mutants would be analogous to a ninety year old who looks and feels forty-five. On this basis we begin to think of ageing as a disease that can be cured, or at least, postponed."

But most of the VCs they approached remained skeptical. According to Guarente, they were fixated on the fact that aging isn't recognized as a disease, hence thought that gerontogene research couldn't possibly yield FDA-approvable drugs. Eventually, though, the scientists hit a resonant chord with Jonathan Fleming, a VC at Oxford Bioscience Partners, which had helped set up Geron. In December 2000, ARCH, Oxford, and a VC firm in the Seattle area invested $8.5 million to get Guarente and Kenyon's new start-up, Elixir Pharmaceuticals, under way.

During Elixir's early days, Guarente served as a director, a role that reminded him of what it had felt like to be a gritty kid from Revere thrust into the elite private school scene. At board meetings, he drily commented in his memoir, "the business people arrive at the table and promptly display their cell phones, pagers, and E-mail devices, the last of which are preferably Blackberrys. Lacking as colorful a plume, I display my Bic pen and occasionally, for variety, a pack of lozenges." After dispensing with Elixir's latest issues, "we leave the room feeling good, and, if the stars are aligned properly, proceed to a fine restaurant for dinner."

But he was no slouch when it came to the business at hand. One of his first moves was to court a seasoned manager, L. Edward Cannon, who had earlier cofounded a biotech named Dyax Corp., to become Elixir's first CEO. Cannon soon hired another biotech veteran, Peter DiStefano, formerly senior director of neurobiology at Millennium Pharmaceuticals, to head research and development. As the Cambridge, Massachusetts, start-up gained momentum, it strengthened its reputation as the leading biotech on aging by adding intellectual property from other prominent gerontologists, such as University of Connecticut researcher Stephen Helfand, discoverer of a fruit fly gerontogene called INDY, for "I'm Not Dead Yet"—a reference to a Monty Python skit.

In early 2003, Cannon arranged Elixir's merger with Centagenetix, the Boston start-up that was pursuing human genes that abet long life. Cannon remained as CEO, and, in conjunction with the merger, VCs who had originally backed the two biotechs committed an additional $17 million to the combined company. A few weeks after the merger, *Boston* magazine ran a lengthy feature on Elixir, headlined TAKE THIS PILL AND LIVE FOREVER, in which Oxford's Fleming was quoted as saying that while the company was "far-out" and "risky," his firm expected to get back ten times its investment in it.

So much for the obscurity of CR-mimetic R&D.

9

RED WINE'S ENIGMATIC DIRTY DRUG

GIVEN THAT SCIENTISTS generally saw the quest for anti-aging drugs as a shady enterprise until around 2000, it isn't surprising that many of the researchers drawn to it before then hailed from the rambunctious end of the psychic spectrum. I mean the kind of person who, during his boy scientist days, didn't patiently wait for the miracle of metamorphosis to play out in his jar of tadpoles, but instead told his little brother that the jar was full of lemonade and watched with fascination as the tyke eagerly gulped its contents down. Or who investigated the power of the compound bow by shooting arrows straight up, then madly scrambling for cover after losing sight of them. Or who figured out how to trap a cloud of methane in a sink during high school chem class, then tossed in a match to see how high the flames would go. In short, the enterprise has been a magnet for scientists like David Sinclair, from whose life these episodes were drawn.*

---

*Some other gerontologists I know did such things as youths. For instance, Steve Austad, the former lion trainer from chapter 3, told me that at age eleven he also liked to shoot arrows straight up. He and his pals did it at night, however, and didn't run for cover—they wanted to get the thrill of hearing the thwack of returning arrows hitting nearby. No one got hit that way, but his archery adventures ended after they switched to playing bow-and-arrow dodgeball and he wound up with an arrow buried in his thigh—that was difficult to explain to his parents, he said.

Sinclair's teachers could readily trace his bent for science, which he excelled in when not clowning around. Both of his parents worked for a medical diagnostics firm in Sydney, and his family's dinner table conversations often touched on topics like the wonderfully grisly effects of certain diseases. To fathom his puckish side, though, one needed to know about his grandmother, Vera, who often oversaw the young "Dr. Sinclair," as she playfully called him, during his formative years while his parents were at work.

The free-spirited daughter of a Hungarian movie director, Vera had given birth to Sinclair's father, Andrew, in her midteens. During the Nazi occupation of Hungary in the 1940s, her family had thumbed their noses at Hitler by secretly sheltering Jewish friends in their apartment. When Soviet tanks rolled into Hungary in 1956, she was one of the daring few who got out, dodging searchlights one snowy night while running across the border to Austria with Andrew, then sixteen. She set up a new life in Australia but continued to flout authority, once getting run off a Sydney beach for wearing a bikini before skimpy was in. Later she went off to New Guinea for a year, where she mingled with natives whose funeral rites included eating the brains of the recently departed. Sinclair proudly recounts that "my grandmother was a former accountant who lived with human flesh eaters."

By his twenties, Sinclair had become a rebel with a cause: making a major discovery in molecular biology. After his postdoc in Guarente's lab at MIT, he landed a coveted post as an assistant professor at Harvard Medical School. Once ensconced in his own lab, he began challenging elder scientists' views, beginning with one of Guarente's theories. At a meeting in 2002, Sinclair questioned his former mentor's hypothesis about how SIR2 is activated and proposed an alternative mechanism that he later detailed in a research report. It was a technical dispute of limited significance—they later agreed that both mechanisms may come into play. But then a weightier contest took shape between them.

In mid-2003, Sinclair's lab made its first big splash by identifying resveratrol and related plant compounds as SIR2 boosters. The discovery promised to lead to CR mimetics targeting SIR2—resveratrol

itself was obviously a candidate. It was a major coup, one that Guarente had hoped to make in his MIT lab or to see happen at Elixir,
which he'd cofounded three years earlier.

Sinclair followed up by forming his own biotech, Sirtris Pharmaceuticals, that aimed to beat Elixir at its own game. Pulling no punches,
the brash Harvard researcher told *Science* magazine in early 2004 that
Elixir "is doing exactly what we're doing, and it's a race." Guarente,
who earlier had tried to enlist Sinclair as an adviser to Elixir, acknowledged in the same article that the contretemps with Sinclair "has run
me through so many emotions, some of which I didn't know I had."
Despite being attacked, one of his feelings was the pride a father feels
when his unruly offspring does something great.

· · ·

The discovery that resveratrol may be a CR mimetic represented a
landmark in several ways, one of which got little notice: The fact that
the advance came out of Harvard Medical School effectively stamped
the establishment's imprimatur on the quest for a way to extend life
span. No institution could have done that more definitively. Stroll
along Boston's Longwood Avenue and you can't help but notice how
the neoclassical white marble buildings at the medical school's center
look down with the august repose of a latter-day Parthenon, as if to
let passing mortals know that Asclepius, the Greek god of medicine,
would be on the faculty except for a minor oversight by the Fates.
Making up for their baffling mistake, the school has brought forth
a series of medical demigods—its professors have racked up no less
than thirteen Nobel Prizes.

Behind the stately exterior, of course, you'll find academe's usual
clashing of massive egos and desperate clawing for tenure. The passionate desires to unravel mysteries and to alleviate suffering are part
of the picture too. Such discord, ambition, and compassion might be
found at any medical school. But only one of them is Harvard's, and if
you make your mark here it tends to loom large and last long.

The study that elevated both resveratrol and Sinclair to stardom
didn't start at Harvard, though. Its first step was taken at Biomol

Research Laboratories, a supplier of research reagents in Plymouth Meeting, Pennsylvania, near Philadelphia. (Biomol is now a unit of Enzo Biochem.) Konrad Howitz, Biomol's director of molecular biology, had developed a test for measuring how fast SIRT1, SIR2's mammalian counterpart, shaves acetyl groups off proteins. The assay was designed to help find compounds that affect the reaction's speed, a sign that they abet or hinder the SIRT1 enzyme's activity. Sinclair had assisted Biomol by providing Howitz with genetic material for making the assay, and in return Howitz had sent him Biomol's finished test kit.

After refining the assay in early 2003, Howitz screened various chemicals with it and turned up two that seemed to boost SIRT1. That was quite unexpected. Enzymes are known as nature's fine-tuned molecular Porsches. Chemical additives that make them run faster than they normally do are very rare. Indeed, pharmaceutical developers have never shown much interest in trying to develop such additives as drugs, for even if they existed, which was deemed questionable for most enzymes, it seemed prohibitively hard to find them.

But Howitz's surprising discovery seemed real, partly because there was something of pattern to it. The two chemicals that seemed to grease SIRT1's wheels, piceatannol and quercetin, are structurally related members of a family of molecules, called polyphenols, that are found in fruits and other plant foods. Polyphenols are antioxidants and had long been thought to confer health benefits by mopping up free radicals. Howitz tested other polyphenols and soon uncovered more than a dozen that sped up SIRT1 in the test tube. The most potent one was resveratrol, a compound found in grape skins, peanuts, and other foods, as well as in giant knotweed, a plant used in traditional Chinese medicine. (The latter is the main source of resveratrol used to make dietary supplements.)

Sinclair could hardly contain his excitement when Howitz told him about the findings—SIRT1 stimulators, he knew, might act as CR mimetics in mammals, possibly including humans. And it didn't take him long to discover that a lot was already known about resveratrol indicating that it could confer a remarkably broad array of health benefits, which is just what a CR mimetic should do.

First isolated in 1940 from the roots of a plant called white hellebore, resveratrol had come to many researchers' attention in 1992 when two Cornell University food scientists suggested that its presence in red wine is what gives the drink, in moderation, its ability to reduce the risk of heart disease. Red wine's cardiac benefits had been spotlighted a few months earlier when *60 Minutes* had covered the "French paradox," a term referring to the curiously low mortality from heart disease in France despite buttery diets there loaded with fat. The popular news show had singled out consumption of red wine as the paradox's most likely explanation.

The Cornell scientists speculated that resveratrol confers heart benefits by lowering blood levels of cholesterol and triglycerides. That idea didn't hold up, though, and in 1997 researchers at the University of Illinois at Chicago proposed a sounder theory: that resveratrol has potent anti-inflammatory effects. The Chicago scientists also discovered that it has a striking ability to block cancer. In fact, when they topically administered resveratrol to mice along with chemicals that induce skin cancer, the number of rodents that got tumors was cut by nearly 90 percent. And resveratrol seemed totally nontoxic.

After the Chicago report, the number of studies citing resveratrol literally rose at an exponential rate. By the time Howitz discovered that resveratrol stimulates SIRT1, there was preliminary evidence that it can lower risk of multiple forms of cancer, Alzheimer's disease, heart disease, strokes, hearing loss, and osteoarthritis. To Sinclair and Howitz, the next step was glaringly obvious: test whether resveratrol can extend life span.

Dousing short-lived yeast cells with the compound was the fastest way to do that, and Sinclair quickly lined up a group of mother cells for the job. He had reason to hurry. Howitz had told him that Biomol had supplied its SIRT1 assay to Elixir. Thus, it was entirely possible that the biotech company had also discovered SIRT1 activators, possibly including resveratrol, and was about to scoop Sinclair's team, which Howitz had joined for the yeast study. Determined to get the experiment right the first time, Sinclair took the unusual step of carrying it out himself rather than entrusting the nitty-gritty work to one of his lab's junior researchers.

The task took him back to his days of doing marathon life-span studies in Guarente's lab, when he often felt like a midwife during a birthing process that went on for a week. After spending most of the day at his Harvard lab plucking away newly budded daughter cells under the microscope, he told me, "I'd drive home with the four plates [containing the mother cells], get home, do a round of plucking at a little lab I set up on the dining room table, have dinner, then do another round. I'd have to tell my one-year-old daughter not to run across the floor, because it would shake my dissecting needle. Sandra [Luikenhuis, his wife] got pretty annoyed."

But she could also sympathize, having experienced the mesmerizing effect of the big experiment herself. A German native, she'd met Sinclair in Australia while studying abroad. After joining him in the United States, she earned a Ph.D. in biology at MIT and then launched her career handling corporate development at a fledgling biotech in Cambridge, Massachusetts. Naturally, she was always the first to hear how her husband's mothers were doing.

When it became clear one fine spring day that yeast cells on resveratrol were replicating significantly more times than control cells that hadn't been dosed with it, Sinclair rushed out of his lab and drove to a park where Sandra was pushing the couple's oldest daughter in a stroller. After spotting her and rushing up, he blurted out, "I think it worked!" A matter-of-fact stickler for details, she calmly asked him exactly what did he mean by that.

No one realized it at the time, but one thing it meant was that Sinclair would soon get more public attention in a single day than the vast majority of scientists get in a lifetime. In August 2003, *Nature* published the study, which showed that adding resveratrol to yeast cells' medium increased their average life span by a spectacular 70 percent and boosted maximum life span by nearly as much. The paper also reported that resveratrol activated the SIRT1 enzyme in human cells, fortifying them against DNA-damaging radiation. And in a tantalizing hint of things to come, it disclosed that resveratrol had extended life span in preliminary studies with nematodes and fruit flies. Concluding with a flourish, Sinclair and colleagues proposed that polyphenols like resveratrol are CR mimetics that can confer many health benefits by activating SIR2-like genes, collectively known as sirtuins (pronounced sir-TWO-ins).

The presence of a possible CR mimetic in red wine put new zing into the French paradox, and the idea that you might brake aging by hoisting glasses of wine instantly became one of the top, feel-good medical stories of the year. Mobbed by the press, Sinclair proved himself a natural at the art of sounding thoughtfully optimistic. "We're making history," he told *Newsweek*. But "I don't think we'll see any Methuselahs in our lifetime . . . we might each get another five years of life." Or we might not, he cautiously mused in a *Boston Globe* interview. In fact, "we clearly don't know if this will work in anything more complex than a fly."

But much of the coverage bordered on the tipsy. The headline in Sinclair's hometown paper, Sydney's *Daily Telegraph*, nicely captured the rosy glow: BETTER RED THAN DEAD, STUDY FINDS.

• • •

Nutraceutical makers wasted no time introducing resveratrol pills that presumably would do for people what the Harvard group had done for yeast. One of the first to play up resveratrol as a possible CR mimetic was Bill Sardi, a writer of consumer health books who interviewed Sinclair at Harvard soon after the yeast study was completed. There were already a number of resveratrol supplements on the market at that point, but Sardi had heard that most of them contained little or no intact molecules of the compound, which quickly degrades when exposed to air. Seizing the day, he organized a company to sell pills that reportedly contain resveratrol whose chemical integrity is maintained by a special encapsulation technology and other tricks—the brand name is Longevinex.

Sardi briefly enlisted Sinclair as a paid consultant, but the arrangement didn't last long. After a dustup over Longevinex's use of Sinclair-attributed quotes on its Web site, the Harvard researcher parted ways with the nutraceutical maker in late 2003 and went on to cofound Sirtris, which planned to develop FDA-approved drugs instead of dietary supplements.

To many scientists, however, the excitement about resveratrol, and especially the buzz about its potential to slow human aging, seemed somewhere between naïve enthusiasm and reckless hype. For one thing, Sinclair's findings hadn't yet been corroborated by other researchers. And even if they held up, it wasn't clear that resveratrol's effects on yeast could be extrapolated to the much more complex process of human aging.

And then there was the lingering two-part question that arose when resveratrol was first linked to the French paradox in 1992: What exactly does the compound do inside cells, and can that mechanism, whatever it is, really explain the benefits attributed to resveratrol?

Unlike the kind of experimental drugs pharmaceutical developers generally prefer to work with, which single out and primarily bind to a single, well-understood molecular target in the body, resveratrol promiscuously consorts with a confusing welter of targets—it's a "dirty drug," as researchers say. That made its mechanism of action extremely difficult to sort out. The fact that it's an antioxidant, for example, might mean that it had extended yeast life span in Sinclair's study by blocking free radicals, not by stimulating SIRT1. That would be nice, but it wouldn't represent a potential breakthrough in human medicine, nor would it support the idea that resveratrol is a CR mimetic. As we saw in chapter 2, antioxidants have repeatedly failed to slow aging in mammals.

To be sure, Sinclair's group addressed the dirty-drug issue in their study. For instance, they reported that resveratrol failed to extend life span in yeast cells with disabled SIR2 genes—evidence that the compound's anti-aging effect was mainly exerted via SIR2.

They also presented a clever evolutionary argument to support the idea that resveratrol is a CR mimetic, and that sirtuins are a major conduit of its anti-aging effect. The theory rested upon the fact that plants churn out resveratrol and other polyphenols when stressed by things like nutrient deprivation, dehydration, and diseases. Indeed, wines with the highest levels of resveratrol are typically made from grapes grown in cool, moist areas in which fungal infections often occur. Thus, plant-eating animals may have evolved inner systems to register the presence of such stress-induced compounds in their food in order to get an early warning of imminent food shortages. The animals could then preventively gear up their CR-like stress defenses before hard times hit, perhaps boosting their chances of surviving. According to this "xenohormesis" theory, sirtuins represent the core of the early warning system—they trigger anti-aging effects when exposed to stress-induced plant compounds, as well as when activated by CR.

But the skeptics wanted confirmatory data, not theorizing. And soon after the famous yeast study appeared, the two most prominent skeptics, Brian Kennedy and Matt Kaeberlein, reported that they had tried to confirm Sinclair's findings but had gotten totally different results.

Their names may ring an ironic bell. Kennedy had helped initiate the Guarente lab's aging research a decade earlier, and Kaeberlein, another alumnus of the lab, had played a leading role in singling out SIR2 as a yeast gerontogene. After leaving MIT, they'd both wound up launching academic careers at the University of Washington in Seattle, where they'd joined forces to conduct further research on anti-aging genes. But a number of their experiments at the West Coast school had cast doubt on some of the Guarente lab's studies, particularly ones indicating that CR's effects are mainly channeled through SIR2. Thus, the Seattle researchers were also skeptical about the related research by Sinclair indicating that resveratrol mimics CR by stimulating SIR2. Not surprisingly, they decided to investigate the compound's effect on yeast themselves.

The results didn't make for chummy Guarente lab reunions. In a 2005 paper coauthored by Elixir's Peter DiStefano, Kennedy and Kaeberlein reported that resveratrol failed to extend life span in three different yeast strains. Further, only one of several different test-tube assays of SIRT1 activity that they employed indicated that resveratrol stimulated SIRT1—the single assay that yielded data like Sinclair's happened to be the Biomol one that the Harvard group had primarily used.

The Seattle researchers attributed the discrepancy between their test-tube findings and Sinclair's to a special molecule incorporated into Biomol's assay. Called Fluor de Lys, it generates a fluorescent glow when the SIRT1 enzyme removes acetyl groups from proteins, providing a readout of the enzyme's activity. Their data suggested that resveratrol stimulates the enzyme only in the presence of Fluor de Lys. In other words, a reagent that was supposed to act as a kind of uninvolved witness to SIRT1-related chemical reactions in the test tube seemed to have gotten mixed up in the reactions as a major player. Kaeberlein and Kennedy concluded that resveratrol's reported SIR-boosting effect was probably a test-tube "artifact" rather than a general phenomenon that occurs in living cells.

Their report appeared in tandem with a similar study led by University of Wisconsin biochemist John Denu. His analysis indicated that various kinds of fluorescing molecules—not just Fluor de Lys—somehow combine forces with resveratrol to amplify the SIRT1 enzyme's action in the test tube. But Denu, who later served on Sirtris's scientific advisory board, was less starkly skeptical about Sinclair's take on resveratrol. In fact, he proposed that unidentified natural molecules might interact with resveratrol and SIRT1 in cells in the same way that the fluorescing molecules do in the test tube. That implied Sinclair's test-tube results weren't just an artifact—surprisingly, they might actually emulate what goes on in cells.

But the Seattle team's critiques went way beyond the minutiae of test-tube assays. Their doubts largely stemmed from studies with various strains of yeast that indicated SIR2's purported link to CR is tenuous at best. For instance, in one strain they studied, CR boosted life span even when SIR2 genes were disabled, contradicting Guarente's hypothesis that CR's effects are primarily channeled through SIR2 in yeast. After conducting many such experiments with different strains, they concluded that the effects of CR and SIR2 are channeled through different, "parallel" biochemical pathways in yeast and, perhaps, animals too. Thus, in their view even if resveratrol did boost sirtuin genes in yeast and other organisms, it probably wouldn't elicit CR's anti-aging magic.

Neither Guarente nor Sinclair was fazed by the attacks. By 2004 they had separately moved on from the study of yeast aging—the main subject of their arguments with the Seattle scientists—to investigate sirtuins in animals. Sinclair collaborated with Brown University's

Marc Tatar and the University of Connecticut's Stephen Helfand in a study showing that resveratrol can modestly extend the lives of fruit flies and nematodes by stimulating their SIR2-like genes.

Meanwhile, Guarente jumped to mice. His lab reported that the tendency of calorie-restricted rodents to become more active, as if impelled to forage more, is absent in mice with SIRT1 deficiencies—an indication that SIRT1 mediates CR effects in the rodents. In another study, his group showed that SIRT1 comes into play in calorie-restricted mice to help suppress formation of new fat cells and to draw down fat reserves. The MIT group also reported that resveratrol, acting via SIRT1, had similar effects on fat reserves—a sign that Guarente and Sinclair were forming a common front against their University of Washington critics.

Important support for the Guarente/Sinclair side came from experiments in various labs suggesting that one of SIRT1's main functions is to help cells stay alive when stressed. A collaborative study by Guarente and Columbia University's Wei Gu in 2001 had raised that possibility, and three years later a group at Washington University in St. Louis led by Jeffrey Milbrandt supported it by showing that SIRT1 helps prevent neurons from withering after nervous-system injury—an effect reportedly abetted by resveratrol. Researchers at the New Jersey School of Medicine in Newark added a study showing that boosting SIRT1 in the heart cells of mice shields their cardiac tissues from free radical damage.

Molecular details of the protective effect were elucidated in another study, coauthored by Sinclair, showing that SIRT1 inhibits genes that sometimes push stressed cells toward suicide. But why, you might ask, would any of our cells, the precious little dears, ever do themselves in? The answer is simple: They're heroically sacrificing themselves to save our lives. The deadly risk springs from the continual damaging of DNA by free radicals and other insults. Such damage is usually repaired, but in some cases it isn't, and on rare occasions it can permanently switch on progrowth genes that lead to the runaway cell division of cancer. To guard against that catastrophe, key genes, most prominently p53 (the two-faced anticancer gene we ran across in chapter 1), are set to induce a kind of orderly self-destruction, called apoptosis, when signs of major damage crop up in cells.

As we age, it seems that many of our cells, increasingly deranged by the ravages of time, wobble ever more perilously along a tightrope between ill-controlled progrowth mode, which tends toward cancer, and hurtful antigrowth mode, which abets aging by robbing cells of their youthful powers of renewal and tips them toward suicide. According to studies led by Guarente, Sinclair, and other scientists, SIRT1 inhibits p53 and other genes that trigger apoptosis. Thus, revving up SIRT1 by means of calorie restriction or CR mimetics might slow aging in part by coaxing stressed cells back from the apoptosis cliff. SIRT1 also has been found to directly induce cells' resistance to stress by enhancing the action of a key stress-response regulator called heat shock factor 1—revving it up is the cellular equivalent of donning body armor.

As all this came into focus, however, Kennedy and Kaeberlein continued to sound a drumbeat of dissonance that reverberated throughout gerontology. The strife seemed to fall somewhere between a Greek tragedy and a lavishly intricate soap opera—a family feud in which two stubborn brothers (Kennedy and Kaeberlein) had turned against their equally headstrong father (Guarente) and brother (Sinclair). The embattled patriarch was characteristically wry when I asked him about the unfolding drama in 2006. "Without naming names, I tell my audiences at conferences that the good news about my lab is that I train students to be iconoclasts," Guarente said. "The bad news is that I'm the nearest icon."

Actually, he didn't need to train Kaeberlein, who, like Sinclair, is a natural at the sweet science of clobbering icons. (Kennedy, the statesmanlike, quietly intense elder brother in the sirtuin saga, seems less comfortable in the contentious role that his findings with Kaeberlein have obliged him to take.) Blunt, bright, headstrong, and driven, Kaeberlein didn't take to his parents' Lutheran faith while growing up in Seattle—in eighth grade he was pointedly absent at his church confirmation ceremony. In high school he aspired to heavy-metal rock fame with pierced ear and long, flowing hair, a look that led to an imbroglio with the basketball coach soon after he joined the team. Free at last after graduating, he put college on the back burner to work the three A.M. shift at United Parcel Service. Three years later, he enrolled in a community college, then decided to go into science after hearing the Darwinian case against creationism laid out in a basic biology course—its rigorous detonation of received wisdoms appealed greatly. After majoring in math and biochemistry at Western Washington University, he went to MIT for graduate studies. He signed up to work in Guarente's lab after attending a seminar on its research—Kaeberlein was especially taken by Sinclair's work, proving yet again that truth is more ironic than fiction.

By 2006, the two sides in the sirtuin dispute were blasting away at each other in acidly worded commentaries, letters to editors, and research reports that purported to correct sins of omission in each other's work. As is often the case in academia, the points of contention—most of which are still subject to debate at this writing—became increasingly arcane as the emotional temperature rose. One of the most hotly contested issues was the proper amount of sugar to give yeast cells in CR experiments. But there was nothing esoteric about the fight's importance—conceivably, both a Nobel Prize and a pharmaceutical fortune hung in the balance.

· · ·

When Clive McCay initiated his pioneering research on CR in the 1920s, his animal of choice for a quick life-span study—the kind that could be carried out by a junior researcher with limited resources—was brook trout. Following in his footsteps, Italian scientists led by Dario Valenzano at the Scuola Normale Superiore in Pisa set out in 2005—two years after Sinclair's high-profile yeast study with resveratrol—to test the compound's anti-aging power in the turquoise killifish. Found in African seasonal ponds that disappear within weeks after rains, the animal lives only about three months.

The researchers tried three doses of resveratrol on different groups of the fish beginning at four weeks of age; in all, 110 fish were fed the substance. Only six weeks later it was evident that the fish on the two higher doses were living longer, and by the end of the study the ones on the highest dose had attained a maximum life span 59 percent greater than the control group had.

Equally arresting, the fish on resveratrol were zipping about in their tanks in a youthful way long after the control animals had lost their get-up-and-go. At nine weeks of age, the resveratrol group's average swimming speed was more than three times greater than that of the controls. The study, reported in February 2006, got little attention in the press. But it was arguably more provocative than anything that had appeared before on resveratrol—it marked the first time the compound's anti-aging promise had been observed on our side of the animal world's great backbone divide.

Mice were obviously next on the list. Sinclair had begun readying a mouse study with resveratrol before the hubbub about his 2003 yeast study had died down. He had discovered a good place to get the rodents: a nondescript, unmarked building tucked away on a side street near Baltimore's waterfront, where the National Institute on Aging maintains a large colony of lab mice for research. He also knew an ideal collaborator for the study: NIA researcher Rafael de Cabo, a burly, convivial native of Spain known for groundbreaking studies on CR and for having a deft touch with rodents. (The latter is not a trivial virtue in biomedical research.) De Cabo had closely tracked the resveratrol story and, though skeptical that the compound would extend life span in mice, was eager to collaborate with Sinclair to test its ability to mimic CR effects in mammals.

But the two were stymied. As a junior scientist living from one smallish grant to the next, Sinclair lacked the twenty thousand dollars needed to pay for the hundreds of rodents required to undertake the ambitious multipart study he designed with de Cabo. Swinging for the fences, they planned to examine the effects of two different doses of the compound on groups of mice fed both high-fat and normal diets—control groups would be required too, of course. Every week that went by with the study on fiscal hold raised the odds that Sinclair would be left in the dust in what was shaping up as the most important research race of his career.

Then one day a man named Tom LoGiudice phoned the Sinclair lab out of the blue to ask about resveratrol. As usual, Sinclair stopped what he was doing to take the call. Unlike many scientists, he genuinely enjoys talking to nonscientists about his work and has a gift for lucid explanation. LoGiudice turned out to be foreman at the U4EA ("euphoria") Ranch, a two-hundred-acre spread near Thousand Oaks, California. An affable Brooklyn native who'd gone west during the Age of Aquarius in an aerodynamically incorrect, hand-crafted double-decker VW van, he had phoned Sinclair on behalf of the ranch's owner, Hank Rasnow. A reserved eighty-five-year-old, Rasnow was thinking about taking resveratrol pills and wanted to know more about their pros and cons.

While answering LoGiudice's questions, Sinclair mentioned his frustrated desire to test the compound in mice. A few days later, LoGiudice, who had a lively personal interest in resveratrol (regular doses of it even livened up his old dog, he told me), arranged for his boss to talk directly with Sinclair. Although Rasnow had never funded research before, he was so taken by the young Harvard scientist that he mailed him a check for twenty thousand dollars immediately after hanging up. Rasnow explained to me in an interview that he had "an eighty-five-year-old passion for longevity, and David sounded like he was really onto something."

Some months later, Sinclair got another call from the U4EA's enterprising foreman. LoGiudice told him that Rasnow happened to know businessman Paul Glenn, a longtime supporter of aging research, and was prepared to help Sinclair approach Glenn for a major grant. Sinclair jumped on a flight to California. Over an impromptu lunch served on paper plates at a condo Rasnow owned in Ventura Beach, the young scientist told Glenn about his dream of setting up a major biogerontology center at Harvard. An alumnus of Harvard Law School, Glenn liked the idea, and in early 2005 the Glenn Foundation for Medical Research in Santa Barbara, California, awarded $5 million to Harvard Medical School to launch a center on the biology of aging, with Sinclair as its founding director.

Meanwhile, Sinclair pushed ahead with the crucial mouse study. He tapped an energetic postdoc named Joseph Baur to coordinate its twenty-seven authors, including experts on rodent pathology, biomarkers, genetics, statistics, and calorie restriction. The mice available for the study were a year old when it got under way—middle-aged for a mouse. Initiating CR at that age typically yields only modest life-span extension at best, and thus the researchers were running a sizable risk that even if resveratrol could mimic CR in mice, their study might fail to show significant anti-aging effects. But using mice whose lives were nearly half over made for a relatively short life-span study. And that was important: Word came through the grapevine that a rival team led by a French scientist named Johan Auwerx had launched a similar study.

To the Harvard group's relief, their mice quickly began generating publishable results. Within three months there were clear signs that when put on fattening diets, mice on resveratrol tended to live longer than undosed control animals. The scientists assembled a paper on the study's initial findings when the mice were about 2.2 years old, at which point resveratrol had cut the mortality risk from the high-fat diet by 31 percent.

Resveratrol also induced a number of CR-like effects. For instance, even though the overfed mice on resveratrol were obese, their livers and hearts appeared to be in really good shape—even better than the organs of normal-weight control mice on regular diets. Resveratrol also prevented glucose levels and insulin sensitivity from trending toward diabetes in mice on the fattening diet. And a gene-chip analysis revealed that resveratrol emulated CR-like changes in nineteen of thirty-six metabolic pathways while opposing 94 percent of the changes associated with the high-fat diet.

Sinclair was suddenly famous all over again—echoes of the study were still resounding two years after its publication in November 2006 when Barbara Walters, glass of red wine in hand, interviewed him about his work for a special ABC report on anti-aging research. To be sure, very high doses of resveratrol were required to induce the CR-like effects. Equaling it in an average-sized person was estimated to require the daily consumption of three hundred glasses of wine. But that didn't lessen the buzz. The New York Times headlined its front-page story on the study, YES, RED WINE HOLDS ANSWER. CHECK DOSAGE. Chip Bok, a cartoonist at the Akron Beacon Journal, comically memorialized the study by picturing a rodent connoisseur commenting on the "timid nose" of the stuff he's sipping while a fellow mouse covers his eyes, exclaiming, "I can't believe you're drinking Merlot."

While giving pause to wine drinkers, the dosage issue didn't diminish the study's biological significance. After all, Sirtris, Sinclair's company, was based on the premise that it would probably take SIRT1 activators more powerful than plain resveratrol to deliver the kind of therapeutic punch needed to ameliorate diseases and win FDA approval. And the team at Sirtris was already making fast progress in developing such drugs.

But some of the study's findings did provide fodder for skeptics. For instance, it showed that resveratrol affected CR-related enzyme-making genes besides SIRT1, including a well-known one called AMPK that's known as a key conduit for metformin's health benefits. (Metformin is the diabetes drug that increases insulin sensitivity.) That suggested SIRT1 may be only one of several of resveratrol's molecular targets in cells that switch on the red-wine ingredient's CR-like effects. If that were true, Sirtris's pursuit of SIRT1 activators might yield little of value.

Further, the study didn't establish that resveratrol slows normal aging. Rather, it showed that resveratrol opposes the effects in mice of awful diets in which 60 percent of calories come from fat. (By comparison, about 34 percent of a typical U.S. citizen's calories come from fat, according to federal surveys.) Thus, strictly speaking, the study showed that resveratrol promises to ameliorate certain diseases caused by pigging out, not to retard aging.

But even the skeptics had to concede that something else quite exciting had happened to the mice on resveratrol: Like the killifish that got the compound, they became livelier over time, seeming to defy the usual slowing with age. As mentioned in the prologue, I got a chance to see this striking phenomenon during a visit to the NIA's Baltimore lab in mid-2006, several months before the study was published.

As I watched, Kevin Pearson, a scientist at the lab, slowly lowered two seriously pudgy old mice onto a device resembling a slowly spinning rolling pin—called a rotarod, it's used to measure endurance and motor skill. Both had been on high-fat diets, but one had also been taking resveratrol. When the rodents touched down, they instinctively began walking in place like log-rolling lumberjacks. The spinner soon accelerated, however, forcing them to run harder until they maxed out and fell harmlessly onto a plastic tray below. Pathetically trembling with exertion, the nonresveratrol mouse dropped after 81 seconds.

But the resveratrol mouse was still going strong at 100 seconds, at 120 seconds, 130, 140. When it finally fell onto the tray after 144 seconds, I was playing the theme song from *Chariots of Fire* in my head. Even fit, young mice rarely last that long. And the rotarod performance gap had widened during the course of the study between the dosed and undosed mice, indicating that resveratrol had had a cumulative benefit. In fact, mice that had been on resveratrol for a year could stay on the rod nearly twice as long, on average, as the control animals.

Two weeks after the *Nature* paper on resveratrol came out, Auwerx's group in Strasbourg, France—the competition Sinclair had worried about—followed up with a report that confirmed and illuminated this invigorating effect. The French researchers had put their mice on doses up to eighteen times higher than Sinclair's group had. Again, the compound was shown to make mice immune to many of the deleterious effects of high-fat diets. But Auwerx's very high doses induced an additional effect: Mice on resveratrol could chow down on rich diets without getting fat. The animals also could run about twice as far on a treadmill as counterparts that weren't on resveratrol. Under the microscope, their muscle fibers appeared to have been remodeled by resveratrol in the same way that they would have been by lots of exercise.

Importantly, the French team's data indicated that the performance enhancement sprang from resveratrol's stimulation of SIRT1 rather than one of its other effects in cells—very good news for Sirtris, which had helped to fund the study. Let's briefly zoom in on this SIRT1-mediated supermouse effect, which might explain a lot of things about resveratrol's effects in us mammals.

Auwerx's group showed that resveratrol indirectly activates an enzyme called PGC-1-alpha, which carries out the heavy lifting of the performance enhancement. (Specifically, resveratrol revs up SIRT1, which in turn stimulates PGC-1-alpha.) PGC-1-alpha had entered the sirtuin picture a year earlier in experiments that had riveted attention on SIRT1's effects on mitochondria, cells' power plants. This line of research, spearheaded by Harvard Medical School's Pere Puigserver, a Sirtris adviser, revealed that during calorie restriction, PGC-1-alpha, juiced up by SIRT1, engenders formation of new mitochondria in muscles. This chain of events probably explains why mice on high doses of resveratrol became elite rodent athletes. Boosting PGC-1-alpha also induces mitochondria to burn stored fat instead of glucose, which may have contributed to the ability of Auwerx's resveratrol mice to eat rich diets without getting fat.

Intriguingly, Auwerx's group also reported that they had linked certain variants of the human SIRT1 gene to a high metabolic rate in people, possibly protecting those carrying such variants from at least some of the deleterious effects of rich diets—a kind of natural version of what high doses of resveratrol do in mice.

Might such people also be endowed with the athletic potential of resveratrol-dosed mighty mice? And could weekend warriors become elite athletes by ingesting large doses of resveratrol?

At this writing, there are no good answers to these questions. But I wouldn't be surprised to find out that resveratrol can have performance-enhancement power in people. Just before the two mouse studies appeared, I was talking to Sinclair about the 2006 U.S. Open, where thirty-six-year-old tennis star Andre Agassi ended his career in tears after losing to twenty-five-year-old Benjamin Becker. When I asked him about Agassi's loss to a young guy who wasn't even ranked in the top one hundred players, he grinned and shot back, "Agassi should have been taking resveratrol."

10

RAPAMYCIN AND THE TALE OF TOR

THANKS TO RESVERATROL'S ability to mimic signs of calorie restriction, scientists seemed tantalizingly close to granting humanity's wish for more of the sweet life as the first decade of the twenty-first century wound down. Still, it wasn't known for sure whether the famous red wine ingredient, or any other substance for that matter, could really retard aging in mammals. Heightening the uncertainty, in 2008 Sinclair and colleagues reported that resveratrol did not extend life span in normally fed mice as it had in ones on high-fat diets.

Because the mice were middle-aged when they began getting resveratrol, the study didn't imply that resveratrol had flunked as a CR mimetic. (Even CR itself doesn't necessarily extend life span when started at that age.) And the mice on resveratrol did show a number of CR-like effects, including delayed formation of cataracts, better motor coordination, improved heart function, and shifts in gene activity reminiscent of CR. But the bottom line was sobering: In light of its failure to increase longevity in normally fed mice, Sinclair's team concluded that resveratrol "does not seem to mimic all of the salutary effects" of calorie restriction.

The Reaper was no doubt chuckling to himself. But a few months later, the Grim One was blindsided by a dramatic turn in the anti-aging quest. In mid-2009, three U.S. labs jointly reported in *Nature* that a drug called rapamycin had clearly extended life span in mice in a way that was redolent of CR's action.

The study was part of the National Institute on Aging's Interventions Testing Program, or ITP, which had been quietly testing possible anti-aging agents since 2003. The program was designed to take the anti-aging quest to a new level of rigor—each compound investigated under its auspices is vetted with life-span studies in mice conducted in parallel at three of the nation's top gerontology labs. Thus, the ITP's rapamycin discovery conferred unprecedented believability on the idea that it's possible to extend life span with a drug.

The testing in triplicate was especially fortuitous in the case of rapamycin, for if the drug's effect had been reported by a single lab it would likely have been dismissed as a bizarre fluke. Rapamycin was found to extend life span in mice that were started on the drug at twenty months of age—roughly equivalent to a person about sixty years of age. Even good, old, reliable CR has rarely produced signs of extended life span in mice when initiated after eighteen months of age. Indeed, in one study, rodents' lives were actually shortened when they were put on CR at such advanced ages. Before the ITP's rapamycin study came out, it would have been very hard to find gerontologists willing to bet that a drug could significantly boost longevity when initiated in mammals so close to the ends of their lives. And after decades of waiting for such a development, many of the field's veterans had little hope that it would occur during their lifetimes—much less while they were still young enough to have a reasonable hope of personally benefiting from it.

• • •

Rapamycin's journey to gerontology's frontier was even more circuitous than resveratrol's. It began in 1964 when a swashbuckling surgeon at McGill University in Montreal named Stanley Skoryna initiated a scientific expedition to Easter Island, an isolated trio of ancient volcanoes that juts from the Pacific Ocean twenty-two hundred miles west of Chile. The goal was to study its people and environment before the planned building of an airstrip irreversibly altered the landscape by facilitating tourist travel to the island.

There was nothing pristine about the place, though. Put on the map in 1722 by a Dutch explorer who first sighted the island on Easter Sunday—hence the name—it features some of the worst ecological devastation wrought by humans. The destruction began centuries ago when the island's Polynesian natives razed its forests while creating the scores of massive, surrealistic stone heads (actually torsos) that Easter Island is mainly known for.

Still, the island wasn't devoid of scientific attractions. One Montreal bacteriologist who visited there in the 1960s brought back five thousand samples of soil and water containing microbes of possible interest. A pharmaceutical lab in Montreal owned by Ayerst Labs, at the time a unit of American Home Products, the predecessor of Wyeth, took on the laborious job of analyzing the samples for natural compounds with therapeutic promise.

Ayerst's decision to investigate the samples wasn't unusual. The pharmaceutical industry got seriously interested in dirt during its formative years in the 1940s, when one Selman Waksman, a Russian immigrant to the United States who studied soil bacteria at Rutgers University in New Jersey, showed that the microbes are rich sources of antibiotics—a term he coined. The compounds have long been thought to be chemical weapons evolved by microbes to use against one another in their never-ending war under our feet, although there is now evidence that many of them are primarily signaling molecules that happen to be toxic to certain microbes at high doses. Waksman enlisted his sizable platoon of grad students in a concerted hunt for these chemicals. In 1943 one of his protégés, Albert Schatz, made history by extracting what became known as streptomycin from the soil bacterium *Streptomyces griseus*. Rushing with wartime urgency from lab to clinic, the researchers quickly arranged with Mayo Clinic physicians to test the compound in guinea pigs, and just months later in a young female tuberculosis patient. She was cured, marking the first time TB had been vanquished. Awarded a Nobel Prize in 1952, Waksman is credited with spearheading the discovery of eighteen antibiotics in all—a fabulous pharmaceutical bounty.

In 1972, Ayerst's microbiologists in Montreal isolated an antifungal compound from another species of *Streptomyces* bacteria found in one of the Easter Island samples. Named rapamycin (Easter Island is also known as Rapa Nui), it showed promise in rat studies for treating vaginal yeast infections. In later research led by Suren Sehgal, a native of India who became rapamycin's main champion at Ayerst, the compound was shown to inhibit immune cells from going on a rampage, which eventually led to its development as a drug to block immune rejection of transplanted kidneys and other organs—the FDA approved it for that indication in 1999. In the early 1980s, rapamycin was also discovered to have antitumor effects, but difficulties formulating it in a stable, water-soluble form suitable for administering to patients discouraged its development as a cancer drug. (Derivatives of rapamycin, however, are now being tested in cancer trials, and one, temsirolimus, was approved in 2007 to treat kidney cancer.)

As rapamycin's therapeutic promise materialized, biochemists gradually unraveled how it works. In 1991, a Swiss team led by Michael N. Hall at the University of Basel tracked down the compound's molecular target. Exercising commendable terminological restraint, they dubbed it simply "target of rapamycin," or TOR. (To be precise, Hall's group discovered that rapamycin's target in yeast cells are enzymes made by two related genes, TOR1 and TOR2.) Later research showed that worms, flies, mammals, and even plants possess TOR-like genes. By 2000 it had become clear that one of TOR's main functions is to control the production of proteins in cells. That suggested a tidy explanation of rapamycin's ability to suppress growth of fungal, immune, and tumor cells: By inhibiting TOR, the compound slows protein production to a crawl, rendering cells unable to expand in size and to produce enzymes that orchestrate their dividing process.

But TOR's tale hasn't been tidy at all. During the late 1990s, Japanese scientists discovered that TOR regulates "autophagy," a process by which cells break down their worn and defective components. (The term comes from the Greek for "self-devouring"—it's sometimes referred to as intracellular cannibalism.) TOR was found by other scientists to control genes that boost cells' defenses against free radicals and other stressors. Another, especially intriguing discovery about TOR had to do with food: Switzerland's Hall and others showed that TOR modulates growth-promoting pathways in tandem with the availability of nutrients, particularly the amino acids needed to make proteins. In times of plenty, TOR ramps up growth and causes cells to literally put on weight. When nutrients are scarce, it slows protein manufacturing.

To gerontologists, TOR's nutrient-driven operation had special meaning—it suggested TOR might be part of the control system that induces calorie restriction's effects. One of the first scientists to take note of this possible link to aging was Zelton Dave Sharp at the University of Texas Health Science Center in San Antonio.

Sharp got interested in TOR while pondering whether there's a common anti-aging mechanism at work in long-lived dwarf mice and ones on CR. Reading up on TOR, he was struck by the fact that nutrient levels aren't the only things that impinge on it—growth-promoting hormones such as insulin also help drive its activity. That implied TOR might switch on anti-aging effects in two different scenarios: when nutrient intake drops during CR, and when levels of progrowth hormones fall. The latter, he realized, is precisely what happens in dwarf mice. Thus, TOR seemed the kind of common thread he was looking for—a gene that might well be so central to aging that any intervention capable of extending life span must turn down its activity to get the job done. If Sharp's flash of insight was right, a drug that targeted and suppressed TOR would have very powerful anti-aging effects.

Well, let's see . . . what compound might target the target of rapamycin?

But when Sharp proposed to colleagues in the early 2000s that rapamycin may have potent anti-aging effects, they generally dismissed the idea—the drug was thought to be too toxic to have any chance of boosting life span. They had a good point: Drugs classified as immune suppressors aren't known for having trivial side effects. Undeterred, Sharp launched a study with Andrzej Bartke, the discoverer of the extraordinary longevity of dwarf mice, that showed TOR activity is abnormally low in the tiny rodents—just what you'd expect to see if suppressing TOR slows aging.

Meanwhile, other researchers began investigating TOR's possible link to calorie restriction. In 2003, Swiss researchers reported that suppressing TOR in nematodes more than doubles their life spans. A few months later, a California Institute of Technology study showed that turning down TOR in fruit flies extends their life spans in a manner strikingly reminiscent of CR. The TOR-tweaked insects were even protected from the life-shortening effects of rich diets, which for a fruit fly means going hog wild on concentrated yeast extract.

In late 2005, Kennedy and Kaeberlein, the prominent skeptics about resveratrol's ability to mimic CR, sounded a loud wake-up call to gerontologists about TOR. They reported that while conducting the first comprehensive search for life-span-extending genes in yeast, they had discovered that TOR-related ones popped up like Woody Allen's Zelig.

Before their groundbreaking study, fewer than eighty yeast genes had been checked for their ability to extend life span when disabled by mutations. Using a novel screening technique, the Seattle scientists managed to test more than five hundred more. Of ten new anti-aging genes they found, six fell within TOR's ambit. Taking a closer look at TOR, they found that CR fails to significantly extend life span in yeast with TOR deficiencies, a sign that CR's effects are largely channeled through TOR. Although the two scientists didn't come right out and say so in their report, they strongly implied that it makes more sense to focus on drugs that tweak TOR, rather than ones that target sirtuins, in the quest for CR mimetics.

Let's step back here and view the big picture on TOR and aging that was rapidly coming into focus. As a key nexus between nutrient intake and growth, TOR is well positioned in cells' web of metabolic pathways to do what needs to be done when starvation looms. Topping TOR's to-do list in lean times is curtailing protein production. That both conserves scarce building materials and cuts cells' energy needs, because manufacturing proteins is one of their most demanding processes. It also frees up resources for use by repair and maintenance systems that underlie the stress response, which is activated when TOR is suppressed, hardening cells against toxic chemicals and other insults.

Intriguingly, rapamycin isn't the only compound that can curtail protein manufacturing via TOR-related pathways. A number of chemical weapons that microbes deploy against foes appear to work mainly by inhibiting protein production, notes Harvard's Gary Ruvkun. Nematodes are among the foes—the worms eat bacteria. That means nematodes often encounter TOR-related toxins, and it seems that young ones are evolved to hunker down in the slow-aging dauer state when they sense that their protein production is being shut down—a telltale sign that they're being poisoned by a bacterial attack. Hunkering down and gearing up their stress response enables them to live long enough to escape the danger zone and develop into reproducing adults. Something similar may occur in mammals exposed to rapamycin. As we'll see, exposing yourself to cups of steaming water laced with that splendid neurotoxic alkaloid caffeine may have a similar TOR-mediated effect.

Here's another benefit of scaling back protein production: It reduces the amount of defective proteins in cells, just as shutting down a car-manufacturing plant curtails its occasional output of lemons. This matters more than you might think. Defective proteins can form toxic aggregates that cells' waste treatment systems can't handle. Indeed, the accumulation of harmful crud in cells has long been thought to play a major role in aging. It's also thought to underlie the massive neuronal die-offs of Alzheimer's and Parkinson's diseases. (Which brings us to an aside on English poet/critic William Empson. Whenever I hear about the crud theory of aging, I find an incantatory line from his 1940 poem "Missing Dates" echoing through my mind: "The waste remains, the waste remains and kills"—yet more evidence, I suppose, that poets are the antennae of the race.)

Stimulating autophagy (the self-eating process that's boosted when TOR is inhibited by CR) also helps prevent crud from piling up in cells. In fact, the idea that crud kills is largely based on studies showing that the amount of autophagy taking place in our cells dwindles as we age—dealing with debris is a costly process, and evolution has geared us to spend inner resources on reproducing, not on keeping the cellular scene tidy during our later years. Lax autophagy apparently contributes to cancer, neurodegenerative diseases, and other scourges of aging.

Calorie restriction's ability to stimulate autophagy via TOR likely confers many health benefits and may well be necessary for CR to slow aging. One of the most important things that amping up autophagy does, it appears, is to help cells get rid of decayed mitochondria, the cellular energy producers that act like smog-belching engines as they wear out—TOR could be seen as overseeing nature's cash-for-clunkers program. Breaking down old cellular components also provides raw materials for recycling into new ones. Cells can even use the break-down products as fuel in times of nutrient scarcity. In fact, newborn mammals rely on autophagy to provide fuel immediately after birth, before their moms start supplying them with milk.

In case you hadn't noticed, most of the things sketched in this portrait of TOR weren't known to be key ingredients of CR's magic before it was linked to TOR. It's too bad that Clive McCay, the back-to-the-land aficionado who discovered CR in the early 1930s, didn't live to witness these developments. He would no doubt have been delighted to learn that so much about CR can be encapsulated by the conserva-tionist's creed: If you want sustainability, reduce, reuse, and recycle.

The picture on TOR and aging is now dominated by the galvanizing mouse study with rapamycin. San Antonio's Sharp deserves credit for instigating it—he formally proposed the study to the overseers of the National Institute on Aging's Interventions Testing Program in early 2004. The ITP itself dates from 1997, when the indefatigable Don-ald Ingram, the former NIA researcher who helped initiate the hunt for CR mimetics, sketched a proposal for the program while sitting around in Sydney's airport—he was waiting for a flight home after attending a gerontology conference in Australia. Ingram envisioned the ITP as a way to test possible anti-aging compounds with more rigor than any single lab had been able to muster—or was ever likely to. The program was also expected to provide the kind of trustworthy data needed to quash dubious nutraceutical fads based on anti-aging claims.

Some of the NIA's leading lights immediately liked the idea, includ-ing Huber Warner, formerly associate director of the institute's Biol-ogy of Aging Program. Now an associate dean at the University of Minnesota, Warner joined forces with Ingram to seek funding for the program. They pitched it to the NIA's top brass in 1998 and a year later worked out the details at a workshop in Bandera, Texas, the self-styled "cowboy capital of the world" and one of gerontologists' favorite conference venues through the years. (They gravitate to Bandera, I suspect, because virtually every one of them is descended from Pecos Bill and Slue-Foot Sue.)

Each year the ITP launches mouse life-span studies with several compounds selected by a steering committee. The same three labs simultaneously test each compound: Richard Miller's at the University of Michigan, David Harrison's at the Jackson Laboratory in Maine, and Randy Strong's at the Barshop Institute at the University of Texas Health Science Center. More than two hundred mice are dosed with each compound, not counting control animals. Importantly, the animals are a genetically heterogeneous variety that Miller developed for aging studies. Using such animals avoids the risk of compiling data that pertain only to a genetically uniform, inbred strain—the kind of oddball mice customarily used in biomedical research. (The genomes of the ITP mice are rather like our own genetic grab bags.) The ITP's primary indicator of an anti-aging effect is the extension of maximum life span. Median life span and aging-related biomarkers, such as insulin levels, are also examined.

The ITP's first round yielded a surprisingly high 50 percent hit rate—two of four compounds showed evidence of enhancing longevity: aspirin and nordihydroguiaretic acid, or NDGA, a creosote-bush extract whose molecules are structurally similar to resveratrol. The results didn't make for big headlines, though. While NDGA and aspirin increased median life span by 12 percent and 8 percent, respectively, they didn't significantly boost maximum life span—an indication that they boost health span but probably don't slow aging. Still, their effects on mice would represent a huge breakthrough if they could be replicated in people. And there are several reasons to think the compounds would have similar human effects. For instance, both happen to combine antioxidant with anti-inflammatory effects. There's also some evidence that both inhibit TOR.

Rapamycin was selected for the ITP's second wave based on Sharp's proposal. (Resveratrol was included in the third set—at this writing results with it aren't available.) Before testing rapamycin, the ITP team struggled for more than a year to find a way to get enough of the drug into mice to significantly inhibit their TOR pathways—the compound doesn't dissolve in water, making it difficult to administer to rodents. San Antonio's Strong finally solved the problem by arranging to have rapamycin particles "microencapsulated" in polymers before they're added to the rodents' food. (Such encapsulation is used with human drugs to enhance blood absorption by preventing their breakdown in the stomach.) By then, however, the mice set aside for the study were twenty months of age, seemingly too old to show significant anti-aging effects. But the researchers figured that going ahead with the study might yield some interesting data. (At least the optimists did—many believed the animals would keel over soon after being put on rapamycin.)

Everyone was astounded by what happened. The stuff worked like magic in all three labs, and their pooled data showed that it had increased maximum life span by 9 percent in males and 14 percent in females. Underscoring rapamycin's remarkable ability to benefit aged animals, male life expectancy at the six-hundred-day mark was increased by 28 percent, while that of females was boosted by 38 percent. The causes of death in the drug-treated and control animals were found to be quite similar, suggesting that the drug didn't merely mitigate a particular disease to increase life span. A follow-up study in which mice began getting rapamycin at a younger age has also yielded preliminary signs of life-span extension.

Unlike mice on CR, the ones on rapamycin didn't lose weight, which means that calling the drug a CR mimetic may be too simple. Still, it's quite possible that a CR mimetic would switch on many of the same anti-aging pathways that CR does without affecting weight. Indeed, a commentary that appeared with the ITP study, written by Kaeberlein and Kennedy at the University of Washington, concluded that it's "a reasonable possibility" that rapamycin acts as a CR mimetic in mice.

• • •

Is it time to start personal anti-aging experiments with resveratrol or rapamycin?

The devil is always in the pharmaceutical details on such questions, and answering this one is particularly tricky because it requires a complex balancing of possible anti-aging effects against possible side effects at a point when there's very little clinical data to guide the decision. And you probably wouldn't get a well-informed answer if you posed it to your doctor—anti-aging research is on very few physicians' radar screens.

Therefore, let us fearlessly grapple with the devil ourselves. Herewith are some details that have jumped out at me while pondering whether to join the early adopters. (David Sinclair, by the way, is on record as taking resveratrol, and another scientist I know told me even before the ITP's mouse study came out that he was planning to take rapamycin as an anti-aging agent.)

At this writing, the astonishing effect of rapamycin in mice speaks louder than any other single finding. Arguably, it represents a more compelling case for taking small doses of the drug than can be mustered for many nutraceuticals on the market. (That's not saying a lot, though.) Of course, rapamycin's long-term effects in people might be quite different from those in mice. But even if rapamycin doesn't slow human aging, it might postpone or avert many diseases of aging. Studies that appeared before the ITP's mouse report suggest that, just for starters, rapamycin can lower the risk of heart disease, bone loss with age, neurodegenerative diseases, and cancer. In a provocative study that's in press at this writing, researchers at the Barshop Institute have shown that rapamycin can slow brain deterioration in mice implanted with genes that induce the neural hallmarks of Alzheimer's disease.

The downsides?

Before delving into them, it's worth noting that several of rapamycin's side effects are also induced by calorie restriction. Indeed, you might say that the drug's apparent ability to mitigate multiple diseases of aging is a CR-like side effect. (The same could be said of resveratrol, by the way.) Also like CR, rapamycin can retard wound healing. That's not surprising, given that inhibiting TOR curbs cells' production of proteins. (But like CR, the drug is thought to preserve latent wound-healing powers in aging animals—stop dosing them with rapamycin and they heal faster than peers.)

A bigger concern is rapamycin's chronic effects on brain function. There's some evidence that the slowdown in protein manufacturing caused by the drug inhibits the synthesis of proteins by neurons. That might hinder formation of new neuronal connections needed to form memories. Rapamycin also can cause blood levels of cholesterol and triglycerides to rise, apparently because it hinders the storage of lipids in fat cells.

Immune suppression is the chief worry with rapamycin. That might increase risk of both infections and cancer—one of the immune system's jobs is to destroy cells that turn cancerous. The mice whose lives were extended by rapamycin were housed in relatively germ-free labs, potentially offsetting the drug's immune risks in a way that people living in the germy world couldn't.

The media stories on the mouse study with rapamycin generally characterized immune suppression as a killer issue for using the drug as an anti-aging agent. But the data on this issue are much less clear-cut than suggested by the knee-jerk classification of rapamycin as a "potent immunosuppressant." Indeed, most clinical studies with rapamycin aren't very informative about its side effects, because the drug is usually given to organ-transplant patients along with other medicines, most importantly cyclosporine, which is a powerful immune suppressant. There is little human data on rapamycin taken alone, and none I'm aware of on its effects when chronically taken in small doses by healthy people.

Consider some findings that don't fit with the idea that rapamycin hammers the immune system. First, many animal and human studies suggest that rapamycin can ward off cancer. (Whether inhibiting TOR suppresses existing tumors is a different matter—cancer trials with rapamycin derivatives have shown mixed results so far.) That's just the opposite of what an immune suppressant is supposed to do.

Rapamycin has been found to suppress formation of proteins that the AIDS virus needs to spread from cell to cell—another strange property for a drug that purportedly hurts immune function. Further, it's likely that the drug's ability to rev up autophagy helps immune

cells break down invading microbes and disseminate their identifying alien molecules like most-wanted posters for use by T and B cells,
key germ fighters. In 2009, low doses of rapamycin were found to
stimulate the formation of "memory T-cells," which patrol the body
for invaders that match the molecular wanted posters—a startling
discovery that suggests the drug, or other TOR inhibitors, might actually be useful for boosting immunity in conjunction with vaccines.

In a 2004 cancer study, very high, repeated intravenous doses of
a slightly altered form of rapamycin caused no "clinically relevant
immunosuppressive effects" or opportunistic infections despite delivering much larger amounts of the drug into the bloodstream than do
the rapamycin pills typically taken by organ-transplant patients. The
most common side effect of the drug, called CCI-779, was acnelike
rashes that, for the most part, spontaneously resolved or were treat-
able with topical steroid cream. The report on the study, which was
funded by Wyeth, CCI-779's maker, concluded that the drug appears
to inhibit TOR at doses "well below" those that cause dose-limiting
toxicity.

In a clinical study reported in 2005, rapamycin was taken orally
for a month by seventy-six heart patients to help prevent proliferation
of cells that can close up arteries implanted with stents. Only three
patients discontinued treatment due to "mild side effects."

Finally, it's important to keep in mind that for all its benefits, CR
itself might be characterized as an immune suppressant. In fact, CR
has been shown to lower white blood cell counts in rodents, monkeys,
and humans by a whopping 20 to 50 percent, and, as noted in chapter
7, it has increased infection risks in animal studies.

There's no doubt that rapamycin can have serious side effects,
especially at high doses. The large doses given to cancer patients, for
instance, sometimes lead to low levels of platelets, which reduce the
ability of blood to clot and can require discontinuance of the drug. But
at low doses, the available data suggest that this is not your father's
immune suppressant—it appears to be more of a nuanced immune
modulator than a blunt instrument.

Here's another thing the media missed in covering the mouse
study with rapamycin: The drug's side effects aren't necessarily tightly

linked to its anti-aging power, and compounds that selectively inhibit parts of the TOR pathway may well be able to slow aging without rapamycin's risks, whatever they are. Scientists are already exploring such selective TOR tweaks. In October 2009, for example, Dominic Withers, David Gems, and colleagues at University College London reported that they'd extended life span in mice by genetically knocking out a TOR-regulated enzyme called S6K1—the altered rodents showed gene-activity changes and preservation of youthful traits much like those induced by CR.

• • •

When mulling resveratrol, I find myself mainly wondering about the extent to which it emulates CR rather than safety.

On the plus side of the ledger, several studies in recent years have suggested that the compound's purported health benefits stem not just from its activation of SIRT1 but also from its stimulation of AMPK, a key enzyme that triggers CR-like metabolic effects when calorie intake drops. AMPK and SIRT1 interact and may reinforce each other's CR-emulating effects, according to a 2009 study coauthored by Sirtris scientists.

Further evidence of resveratrol's ability to mimic CR was reported in 2008 by University of Wisconsin researchers, who showed that resveratrol's gene-activity fingerprint in certain mouse tissues closely resembles that of CR. Guarente's group at MIT has shown that mice implanted with extra SIRT1 genes display reduced glucose and insulin levels, as well as other metabolic peculiarities reminiscent of CR. In a similar study, Columbia University researchers found that mice equipped with extra SIRT1 genes are protected against the normal tendency toward diabetes that often occurs in aging rodents.

But some scientists still doubt that resveratrol deserves to be classed as a CR mimetic. They cite its failure to extend life span in normally fed mice, as well as a 2007 study by Linda Partridge and colleagues at University College London showing that it didn't extend life span in fruit flies, and did so only inconsistently in worms—findings at odds with earlier data reported by Sinclair and colleagues. The fact that rapamycin has been shown to extend life span in normal, healthy mammals—and that so far resveratrol hasn't—figures prominently in the skeptics' thinking.

The skeptics also cite growing evidence that resveratrol doesn't directly stimulate SIRT1: test-tube studies in late 2009 and early 2010 by scientists at Amgen and Pfizer, respectively, echoed the earlier ones suggesting that an experimental artifact involving fluorescently tagged molecules has given the red-wine compound an undeserved reputation as a direct SIRT1 activator. The studies, however, left open the possibility that resveratrol boosts SIRT1's activity indirectly via some sort of chain reaction in cells.

Still, I find the rodent and other animal data on resveratrol more compelling than such test-tube studies, and the compound has repeatedly been shown in different labs to induce CR-like effects that are of special importance during the age of pandemic obesity. Resveratrol's ability to mimic CR's effects on blood levels of glucose and insulin, as well as its apparent power to engender new mitochondria, neatly oppose the deleterious effects of rich diets. Resveratrol and other SIRT1 activators appear to reduce the chronic, low-level inflammation that's exacerbated by obesity and is probably a major contributor to diseases of aging. Resveratrol seems to have particularly compelling cardiac benefits.

There's also some evidence that resveratrol and rapamycin have overlapping modes of action. In 2007, Sinclair's lab reported that inhibiting TOR (rapamycin's effect) activates SIR2 in yeast while ramping up stress-response genes thought to help slow aging. Toren Finkel and colleagues at the National Heart, Lung, and Blood Institute have found that SIRT1 regulates autophagy in mice, and that rapamycin's stimulation of autophagy appears to be channeled through SIRT1 as well as TOR. Mikhail Blagosklonny, a TOR expert at Roswell Park Cancer Institute in Buffalo, New York, even speculates that resveratrol's anti-aging effects may actually stem from its ability to inhibit TOR, if only indirectly.

But is resveratrol safe?

If the data on rodents given copious doses of resveratrol are any indication, the compound is remarkably benign—especially considering the profound metabolic shifts it can induce. The mice in the Sinclair group's 2006 study got daily doses of 22.4 milligrams per kilogram of weight. Rats on daily resveratrol doses of up to 700 milligrams per kilogram of body weight have reportedly shown no adverse effects. It's hard to say exactly what the equivalent human dose would be, because people and rodents metabolize drugs differently. But it's likely to be thousands of milligrams a day. (Resveratrol supplements currently on the market generally contain tens to hundreds of milligrams.)

One notable risk, however, has repeatedly come up in research on SIRT1, raising concerns about chronically taking large doses of resveratrol: cancer. SIRT1 levels have been found to be highly elevated in several types of tumor cells. Moreover, SIRT1's ability to convey a don't-die signal to cells means that SIRT1 activation by resveratrol or other compounds may sometimes abet cancer by inhibiting cells' suicide programs and preventing them from eliminating themselves when damaged in ways that could lead to unchecked growth. Some scientists believe that SIRT1 inhibitors—antiresveratrols, so to speak—may help fight tumors.

On the other hand, there's a wealth of data suggesting that taking resveratrol fights cancer. Administering resveratrol to mice has been found to reduce tumor formation in multiple studies. A 2008 study coauthored by Guarente and Sinclair, for example, showed that amplifying SIRT1 in mice prone to intestinal cancer suppresses their tumors. Another 2008 study, by NIA researchers, suggested that resveratrol can help treat certain breast cancers. Mice on chronic, high doses of resveratrol have shown no signs of increased cancer. And Sinclair has marshaled evidence that SIRT1's anti-aging magic largely stems from its ability to preserve "genomic stability," warding off DNA damage underlying cancer and other diseases of aging.

When weighing all these pros and cons, I find myself repeatedly circling back to the sage cliché that the dose makes the poison. (Even overdosing on water can be fatal in certain circumstances.) There's no doubt that the dose is also responsible for the anti-aging effect when it comes to CR mimetics. After mulling these dosage issues, I find myself reluctant to join the early adopters until I can get at least partial answers to two questions: What's the minimum dose of rapamycin or resveratrol (or of metformin or mannoheptulose, while we're at it) I'd need to take to elicit various CR-like effects? And does combining these or other putative CR mimetics yield synergistic effects on biomarkers linked to retarded aging? (If the answer to the latter is yes, a multicompound cocktail with very low, safe doses of each might work nicely.)

Truth in advertising: I'm a native of the Show-Me State, Missouri, which has proudly singled out the mule as its emblematic animal. My mulishness has limits, though. As a baby boomer, I'm manifestly running out of time, and the drawbacks of trying a possible CR mimetic seem to be steadily shrinking—while there may be little to gain at my age, there's less to lose with the passing of each day. Thus, while hesitant about joining the pill poppers, I've become an early adopter lite, striving to get items into my diet that probably activate CR-related pathways. As noted earlier, resveratrol is found in many foods, as are related polyphenols thought to activate SIRT1—the list begins with red wine, but it also includes dark chocolate, peanuts, cherries, grape juice, and apples, among many others.

For a rapamycinlike jolt, moderate coffee consumption stands out as an adopter-lite option. Caffeine has been discovered to inhibit TOR, which means drinking a cup of coffee may be akin to taking a little dose of rapamycin. In fact, scientists at the University of Geneva in Switzerland estimate that moderate coffee consumption can reduce a person's TOR activity by about the same amount that extends the life span of yeast cells dosed with rapamycin. Additionally, there are several studies suggesting that aspirin, which many of us already take in low doses to ward off heart attacks, can inhibit TOR.

But when will my two pressing questions about rapamycin and resveratrol be answered?

Unfortunately, addressing them will require clinical tests that few institutions have the resources or motivation to pursue. Someday the National Institute on Aging may expand the ITP to include a clinical-testing arm. But I'm not holding my breath. As the University of Michigan's Richard Miller has observed, "A president who announces a war on cancer wins political points, but a president who publicly committed the government's resources to research on extending people's life span would be deemed certifiable." A few private foundations might fund such studies. So might biotech companies, albeit in the context of developing CR mimetics to treat specific diseases. But the vision, buzz, money, luck, and chutzpah required to make this happen in biotech has really come together only once so far—at Sirtris, the subject of the next chapter.

11

SIRTRIS, MASTER VOYAGER OF THE VORTEX

FACING THE CHARLES River between Harvard and MIT, the Polaroid Corporation's former headquarters in Cambridge, Massachusetts, has the kempt, desolate air of a landmark whose significance has faded from living memory. The techies flashing by its art deco façade en route to the next big thing are too young to have witnessed the tale of hope, genius, and hubris that unfolded here as Edwin Land's instant-photography concern rose to glory in the 1960s, then lost its way in the digital age and tottered into bankruptcy. But the moral of Polaroid's story—never stop looking anxiously over your shoulder—was very much on the mind of Sirtris Pharmaceuticals CEO Christoph Westphal as he strode past Polaroid's vestige in late 2006 on his way to work. As usual, he was commuting by foot from his home a mile and a half away—walking met his inner need to press forward at all times.

Westphal had begun the year with high hopes for negotiating a partnership with a pharmaceutical company that would pump millions of dollars into his biotech for rights to its early-stage medicines. But nothing had come of a series of visits by big pharma envoys during the spring and summer. More worrying, GlaxoSmithKline, the industry player that had shown the most knowledge and excitement about Sirtris's research, had stopped talking to it about a possible deal, suggesting that it planned to compete rather than collaborate. Westphal didn't need to look over his shoulder to see the looming competitive threat. Sirtris had a head start in developing sirtuin-based medicines, but the pursuing giant's long shadow extended well out in front of it.

Asserting its edge, Sirtris had begun clinically testing its first drug less than two years after opening its doors. Dubbed SRT501, the medicine contained resveratrol in a form designed to get more of the compound into the bloodstream than existing dietary supplements did. The biotech had also aggressively ramped up research on novel SIRT1 activators far more potent than resveratrol. But trying to stay ahead of rivals many times its size was turning Sirtris into a cash bonfire, and it needed to raise a lot of money in the near future.

Westphal is a tallish, solidly built man with closely cropped brown hair and a disarming, jocular manner—at first glance you can readily picture him as the wisecracking high school halfback he never was. His quickness in conversation tells a more accurate story. He tends to see where you're headed before you get there, and I've sometimes found that my verbal exchanges with him taking on a staccato, Twitter-like quality, as if we were both frantically trying to extend life on the cheap by packing more into the time allotted.

Bounding into Sirtris at seven-thirty, he walked into the office of the vice president of finance, Paul Brannelly, and immediately began hashing out details of Sirtris's recently hatched plan to file with the Securities and Exchange Commission for an initial public offering of stock. If all went well, the IPO would provide Sirtris with enough capital to demonstrate signs of efficacy in clinical studies with SRT501, a critical milestone. Talking fast, Westphal seemed to have forgotten that he was still wearing his coat and fleece stocking cap. There was something else he'd apparently overlooked, a matter of somewhat greater significance: Sirtris was at an absurdly early stage to go public.

Only a few months earlier an article in the *Wall Street Journal* had highlighted the withering skepticism biotech start-ups were facing when trying to sell stock to the general public. "These days," the article starkly noted, "few investors will even look at a biotech IPO unless the start-up has a drug candidate in late-stage testing, meaning it might be only a few years away from possible regulatory approval." A number of young biotechs had been forced to cancel planned IPOs, and others had accepted miserably little for their shares. When Westphal first told Sirtris's board in mid-2006 that he planned to move quickly toward an IPO, he recalled, "they told me I was absolutely insane."

But he'd always maintained that Sirtris was special, and evidence was mounting that he was right. A month earlier, the famous mouse study led by Sirtris's founding scientist, David Sinclair, had appeared, showing that resveratrol could render the rodents largely immune to the health fallout from rich diets, boost their exercise endurance, and perhaps even mimic CR's anti-aging effect. The resulting hullabaloo had made Sirtris the most celebrated early-stage biotech since the genomics craze of the late 1990s. When I visited Westphal soon after the study was published, he commented that "there's a funny thing about this company—it's the first biotech I've seen where half the board wants to take the drugs it's developing."

Sirtris had another asset that set it apart: Westphal himself. Before cofounding Sirtris in 2004 at age thirty-six, he'd been one of the most prominent young venture capitalists in the country. He'd cobbled together five high-flying tech companies in five years, including two biotechs that had already reached a stunning combined market value of more than $1.3 billion. Given his track record, people tended to pay attention when he described Sirtris as his entrepreneurial pièce de résistance. As he told me in an e-mail, his standard line to potential investors in Sirtris was that he'd quit his meteoric VC career to lead the company because he was convinced that it represented a "once in a lifetime chance to change medicine and improve society." He actually seemed to mean it.

• • •

Westphal showed his entrepreneurial flair at about the same time of life Tom Sawyer did. His parents, natives of Germany who have spent most of their lives in the United States, remember him before age ten as a quiet, bookish youngster. While his toy train set gathered dust, he plowed through history and other nonfiction books, once even reading the dictionary from A to Z, recalled his father, Heiner, a geneticist and lab director at the National Institute of Child Health and Human Development.

But around age eleven his inner capitalist boldly asserted itself. After arranging to mow the lawns of various people in his neighborhood, he enlisted friends to do the work and took a percentage as manager. As a high schooler, he began earning returns on the cello lessons he'd dutifully taken since age six—he's an accomplished player—by organizing a string quartet with friends that played classical music gigs in restaurants and other venues. (It was the first of three quartets he assembled during his student years.) While spending a summer in Germany as a teen, he used three thousand dollars saved from the lawn-mowing business to buy a clutch of popular name-brand suits that he'd heard were scarce in the United States; he shipped them back in his cello case and made a handsome profit. The instrument came home in a cheap bag.

After graduating summa cum laude from Columbia in three years with a triple major in economics, literature, and biology, Westphal went to Africa for nine months on a Rotary Scholarship. Ostensibly working with a medical team and studying in Cameroon, he spent much of the time rambling around the continent as if it were a giant dictionary to be read cover to cover. To his mother, a physician, it was the vicarious year of living dangerously. "He was always very convincing," she drily recalled, "when he told us over the phone that his school in Cameroon was closed for some reason, so he had gone to Chad or some other country that he would describe as this wonderful, safe place. Then we would wait and pray for the next call."

No telling where it would come from—he was in all-out Kerouac mode, cramming life skills and getting, in his words, "really grounded. I was living in rooms with dirt floors, without running water, without electricity, and I was very happy. It helped me understand what the world is about."

He also found himself in "some seriously scary situations. One time I was walking along the bank of the Congo River and these guys came at me with guns drawn, saying they were government agents. They took me into a car, and I thought they were going to kill me. But they just wanted to rob me, and took my money.

"I think traveling around Africa in your early twenties with nothing but a backpack is good preparation for being a venture capitalist. You're on your own and you have two minutes to size up this person who's approaching you in the street at night in Kinshasa. Do they want to kill you, or do they want to help you? What's your judgment?"

Post-Africa, he polished off two Harvard doctorates in near record time, about six years. While getting his Ph.D. under renowned geneticist Philip Leder, he somehow found time to serve as lead author on a half dozen studies, including ones published in prestige journals such as *Cell*. He slogged through an M.D. on the side—he'd once pictured himself leading a life like Albert Schweitzer's but then realized he hated the regimentation of medical practice. The best thing about going to med school, he told me, was that it brought him into contact with his future wife, who at the time was earning a Harvard Ph.D. in biology. It also led him back briefly to Africa, where he worked at the Albert Schweitzer Hospital in Gabon for three months after getting his degree—he got a lot of practice delivering babies, and later he delivered his own kids.

Next he spent two years at management consultant McKinsey & Co., then joined Polaris Venture Partners in Waltham, Massachusetts, to take up his true calling, starting companies. He honed his management skills during his five years at Polaris by serving as founding CEO of four of the companies he launched, including Alnylam, which became one of the hottest biotechs of the early 2000s.

Soon after the publication of Sinclair's 2003 report on resveratrol's life-extending effect in yeast, Westphal met with the Harvard researcher to discuss starting a company. They didn't immediately hit it off. Westphal nearly walked out when the scientist refused to disclose proprietary details of his work.

But Sinclair enjoyed verbally sparring with the VC, who asked all the right questions. Westphal was clearly intrigued by the promise of sirtuin-based CR mimetics. Sinclair's showmanship also impressed him, and he correctly guessed that when it came to selling the sirtuin story to investors, the scientist would pull more than his weight in gold. Although Sinclair had earlier planned to start Sirtris with another group of high-tech entrepreneurs, he soon decided to join forces with Westphal. In spring 2004 the two jointly began making the rounds at Boston-area VC firms soliciting funds for Sirtris's first financing round, which Westphal's VC firm was set to lead.

It was a harder sell than they'd expected. In the wake of the high-tech crash that began in 2000, VCs were more acutely aware than ever that early-stage biotechs face about the same odds Evel Knievel did when revving his Skycycle to jump the Snake River Canyon. (He didn't make it.) It didn't help that Sirtris would be an offshoot of anti-aging research, still widely seen as borderline quackery. "People were telling me this is crazy, this won't work, and we're not sure about Sinclair's science," Westphal told me. Still, some of the VCs they approached were very taken by Westphal's willingness to accept a huge pay cut by quitting Polaris to become Sirtris's full-time CEO—it was a form of walking the talk that the financiers found both utterly compelling and morbidly fascinating.

In August 2004, Sirtris was launched with a $5 million investment from Polaris and three other VC firms. Westphal's mentors at Polaris weren't happy when their star protégé resigned, and because their firm was Sirtris's main backer he felt like a wayward son forced to borrow money from a father he'd seriously pissed off. But he never looked back. He and Sirtris's first few employees set up shop in a Waltham, Massachusetts, office that was "so small we took out the doors because we couldn't shut them and all sit down at the same time," he recalled. "When the weather got cold, we were wearing Polarfleeces, sitting next to space heaters. If our investors had walked in, they would have thought we'd set up a shell company."

As promising preliminary data from Sinclair's resveratrol study in mice rolled in, Sirtris raised $27 million from VCs in early 2005. (That made VCs' total bet on the biotech $45 million, including $13 million in late 2004.) The company moved to larger quarters in the building next to Polaroid's former headquarters. Westphal's new office, which he shared with Sinclair, who served as a scientific adviser and board member, had all the roominess of a walk-in closet. There was a door too.

From the start, Westphal kept Sirtris at a safe remove from the anti-aging quest's shady side. His first step was to add a glittering list of names to Sinclair's as scientific advisers, including Nobel laureate biologist Phillip Sharp (MIT); gene-cloning pioneer Thomas Maniatis (Harvard); world-leading biomedical inventor Robert Langer (MIT); and Thomas Salzmann, formerly executive VP of Merck's research labs. With such eminent scientists on board, he had no trouble hiring prominent academic experts on sirtuins as consultants. One of the few he didn't sign up was MIT's Guarente, who at the time worked with Sirtris's crosstown rival, Elixir Pharmaceuticals.

Westphal also brought in a prestigious board, including hedge fund manager Rich Aldrich, who had helped build Vertex Pharmaceuticals into a biotech powerhouse; M.D.-Ph.D. venture capitalist Stephen Hoffman, founding CEO of Allos Therapeutics; Richard Pops, then CEO of Alkermes, a Cambridge, Massachusetts, biotech; and biologist Paul Schimmel at the Scripps Research Institute, cofounder of half a dozen biotechs.

A number of biotech start-up artists besides Westphal might have lured a similarly impressive set of advisers and backers. But only the rarest of entrepreneurs could have managed to get so many big egos and potentially conflicting agendas pulling together in a single direction with maximum force for nearly four years, as he did. When I asked him how he did it, he grabbed one of the few books he keeps in his office, on abnormal psychology, and loaned it to me. I thought he was joking. But while poring over it, I noticed that it tended to fall open to the pages on narcissistic personality disorder, as if that part had been frequently consulted.

In any case, it's arguable that the main ingredient of Sirtris's early success—which set up its emergence as gerontology's most successful spin-off—was simply that Westphal, for all his relentless drive and formidable intellect (did I mention that he's fluent in four languages?), can charm the socks off a Gila monster. While sitting in on dozens of meetings at Sirtris between 2006 and 2008, I was struck by the fact that he didn't suck all the oxygen out of the room, as many charismatic leaders are wont to do. One reason, it seems, is that he's possessed of, and by, a strong sense of the general irony of things (think of Germans like Günter Grass and Heinrich Böll), which often prompts him to step back and poke fun at himself. When a local TV news team arrived in early 2007 to interview him and others at Sirtris, for instance, he couldn't resist donning a white lab coat and safety glasses before going on camera, a goofy affectation (the CEO hasn't done lab work for years) that caused much mirth at the company. "We're in trouble if we lose the ability to laugh at ourselves," he said when I asked what had come over him.

• • •

By mid-2006 Westphal had installed a set of high-octane managers, and Sirtris was hitting on all cylinders. One of his first hires had been Jill Milne, a dogged, exacting former Pfizer researcher who was leading the hunt for novel SIRT1 activators—the better-than-resveratrol drugs Sirtris hoped to turn into patent-protected medicines. In 2006, she was joined by Mike Jirousek, a lean, intense biochemist who resigned as head of a large drug-discovery group at Pfizer to lead discovery at Sirtris.

Garen Bohlin, an unflappable former Wyeth executive, signed on as chief operating officer. He played Scotty to Westphal's Captain Kirk when it came to the budget. ("I've giv'n her all she's got, Captain, an' I canna give her no more.") Michelle Dipp, a buoyant, Oxford-educated M.D.-Ph.D. from Texas, led business development—known at Sirtris as "General Dipp," she quickly emerged as a Westphal-like wheeler-dealer. Biotech veteran Peter Elliott directed clinical testing. A charming native of Wales whose office abounded with sheep figurines, Elliott was known for codeveloping an important new cancer drug, Velcade, at Millennium Pharmaceuticals.

The fact that a clinical expert of Elliott's stature had joined Sirtris underscored one of its major advantages: It was in a position to test medicines in humans much earlier than most biotech start-ups. Resveratrol, the main ingredient of SRT501, had been ingested by humans for thousands of years with no known ill effects. That didn't guarantee that the large, concentrated doses Sirtris planned to give patients would have no side effects. But it did mean that SRT501 could be classed as a dietary supplement, obviating the need for lengthy preclinical testing before trying it in people.

When Sinclair had first started thinking about Sirtris, he decided to focus initially on developing diabetes drugs—pursuing anti-aging pills was obviously a nonstarter. (Recall that the FDA does not consider aging to be a condition warranting treatment.) Given resveratrol's apparent lack of toxicity, SIRT1 activators seemed likely have fewer side effects than existing diabetes drugs, giving Sirtris a major competitive advantage.

By rapidly initiating a clinical trial with SRT501 in diabetes patients, the company made itself interesting to investors who generally avoided early-stage biotech. That was one reason Westphal was able to turn it into a liquidity-fueled rocket that left rivals behind in the sirtuin space race. But SRT501 was just for starters. The main payload was to be Sirtris's novel SIRT1 activators, which were expected to be more effective than SRT501 and, just as important, to provide Sirtris with "composition of matter" patents that could be licensed to big-pharma partners.*

The biotech was playing an especially high-risk game, though. Even if SIRT1 activation could ameliorate diabetes—a big if—SRT501 might not get enough of the drug into humans' bloodstream to yield signs of efficacy. The resulting clinical failure would probably say little about the general value of sirtuin-based drugs, and Sirtris's more potent SIRT1 activators might well overcome the problem. But a failed "proof of concept" trial with SRT501 could easily scare away investors and send the Sirtris rocket tumbling end over end. And ramping up clinical trials so early would test Westphal's money-raising skill as never before.

Speed and efficiency became a mantra at Sirtris as its burn rate escalated. To help minimize costs, Elliott orchestrated initial diabetes trials with SRT501 in India. (Lest you think that that common industry practice is unethical, consider that many sick people in places like India can't afford any medicines at all, much less therapy given in

---

*Resveratrol is a natural molecule that can't be patented. Sirtris did seek to patent SRT501's special formulation of resveratrol, which reportedly delivers four times as much of the compound into the bloodstream as do widely sold supplements, based on comparable doses. But such patents are relatively easy to circumvent. Thus, SRT501 was of limited value to Sirtris as a bargaining chip in negotiations with big pharma.

stringent trials designed to meet FDA standards.) Saving both time and money, Westphal, Sinclair, and Elliott used themselves as unpaid volunteers for preliminary tests of SRT501's ability to get significant amounts of resveratrol into the bloodstream—Sinclair, who hates needles, often looked ashen after the blood-drawing sessions that were required. The biotech also farmed out much of its chemical synthesis work to a contractor in China.

Meanwhile, Westphal kept the liquid-fuel tanks topped up. Before Sirtris went public in 2007, he brought in a whopping $113.5 million from venture capitalists and other investors in five financings—far more than any previous gerontology spin-off had secured at such an early stage. Sirtris's ability to raise money partly stemmed from growing excitement about anti-aging research in general, and about Sirtris's in particular. But Westphal made use of the buzz with great care. He'd worried from day one that Sirtris would get sucked into what he called "the vortex of inflated expectations" surrounding anti-aging science. On the other hand, he knew that public fascination with Sinclair's anti-aging research could be a powerful money magnet.

Sirtris's need for unassailable scientific respectability predominated at first, and the company's releases before 2006 merely stated that it was focused on sirtuins without mentioning its link to anti-aging research. But figuring out how to safely tap the vortex's energy was never far from Westphal's mind. And as evidence accumulated that Sirtris's underlying science was sound, it seemed less risky to do so.

By mid-2006, Milne's team had found SIRT1 activators that were a thousand times more potent than resveratrol. As Sinclair had expected, and the more guarded Westphal had ardently hoped, the novel compounds engendered CR-like effects in mice. In fact, Sirtris's biologists showed that both SRT501 and the new SIRT1 activators, whose molecules were very different from resveratrol, significantly reduce blood glucose and insulin in mouse models of obesity and diabetes. SRT501 also had been found to block optic nerve damage in mice with a disease akin to multiple sclerosis, indicating that Sirtris's drugs might ameliorate various forms of neural degeneration. There was a collective sigh of relief at the biotech as the encouraging animal data rolled in.

The glow at Sirtris was further stoked by promising, unpublished data flowing from the labs of its academic collaborators. (While building its brain trust, Westphal had nearly cornered the world market on sirtuin expertise.) The parallel studies with resveratrol in mice led by Sinclair and by France's Johan Auwerx were especially heartening. The emerging evidence that SIRT1 activation rejuvenates mammals' mitochondria particularly excited Sirtris's drug developers, because decay of the cellular power stations had been implicated in many diseases of aging, including diabetes. In late 2006, Dipp began setting up a clinical test with SRT501 in patients with MELAS, a rare genetic disease involving defective mitochondria that leads to fatal brain and muscle degeneration.

Ironically, the progress at Sirtris was quietly unfolding at the same time that doubts about Sinclair's discoveries were being raised and forcefully reiterated by the University of Washington's Kennedy and Kaeberlein. Sinclair was frustrated by the fact that the biotech's trade secrecy prevented disclosure of its findings, which represented compelling support for his embattled view that resveratrol and other SIRT1 activators are CR mimetics. Westphal was chafing too. As the CEO well knew, VCs who were considering making bets on Sirtris, as well as potential drug-company partners, were likely to get an earful from Sinclair's critics while doing their due diligence. And the skeptics were essentially arguing that Sirtris's pursuit of SIRT1 activators was folly.

But publicizing the telling results with Sirtris's novel SIRT1 activators in a press release would likely backfire—the critics would trash it as unsubstantiated corporate hype. When Westphal and Sinclair proposed instead that Sirtris report the findings in a peer-reviewed journal, the biotech's board nixed the idea. Landing the report in a top publication like *Nature* would require disclosing the SIRT1 activators' chemical structures. That might enable competitors to rapidly develop variants of the molecules that weren't covered by Sirtris's patents.

But Westphal, who's a good listener but not so good at taking orders, wasn't about to drop the idea. After the board dismissed it, he found himself wondering whether he would get fired if he went ahead.

He had another reason for wanting to stir excitement about Sirtris's progress. Sinclair, who was continually bombarded with requests for advice on taking resveratrol, was convinced that there would be huge demand for SRT501 if it were sold as a dietary supplement. Westphal saw a lot to like about the idea. It would obviate Sirtris's need to deal with the FDA on its first product, cut its cash-burn rate, and make him less beholden to the VCs supplying money to Sirtris, who dominated its board. There were precedents. Italy's Sigma-Tau Pharmaceuticals, for instance, was mainly a prescription-drug company yet also sold clinically vetted nutraceuticals for various diseases.

But even if Sirtris avoided the kind of hype that gives the nutraceutical business a bad name, selling a dietary supplement with purported anti-aging effects would force it to head right for the vortex Westphal had long avoided. The company's eminent scientific advisers, and some of its staff scientists, might even resign, fearing taint by association. Besides, if SRT501 really worked as hoped, offering it as a nutraceutical would probably mean forgoing much greater future returns from its sales as a prescription drug.

After mulling the nutraceutical option for months, Westphal and his top managers decided it was best kept in reserve until more was known about SRT501's prospects as a prescription drug. Nevertheless, the safe distance from the vortex that Sirtris had maintained since its inception was about to end. After the electrifying mouse studies on resveratrol came out in late 2006, Sirtris was vortex-bound whether Westphal liked it or not.

• • •

Sirtris was deluged by calls after the results of Sinclair's study were widely reported in the media. Everyone from reporters in Australia to a pair of elderly sisters planning to add the biotech's drugs to their daily doses of red wine were suddenly trying to get through to Westphal, as were potential new investors. When I stopped by Sirtris soon after the news broke, Westphal's tiny office was more cramped than usual because of several large, handsomely wrapped packages piled near his desk—Christmas gifts, he told me with pained amusement, from people he'd never heard of. He hadn't quite gotten used to his new image as keeper of a golden goose.

He liked its eggs, though, and six weeks after the resveratrol study was published Sirtris took the first step toward an initial public offering at an all-day confab with investment bankers. Leaning back in his chair in the biotech's bare-bones conference room, Westphal watched with judicious absorption as delegations of dark-suited bankers came through, each allotted an hour to make a presentation aimed at convincing him to hire them to manage the IPO. His sober demeanor belied his thoughts—he was ecstatic.

Early-stage biotech start-ups often find themselves playing the role of beggars rather than choosers at such meetings. Big investment banks often don't even bother to bid for the small potatoes represented by tiny biotech firms' IPO fees. Sirtris's meeting, however, had become a jam-packed beauty pageant—with Westphal as the judge. J.P. Morgan even sent its director of equities trading. Struck by his casual air of command, Westphal later referred to him as "the regal guy."

As one of the last teams of bankers filed into the room with smiles and handshakes, Hoffman, a Sirtris director whose sense of humor is similar to Westphal's, leaned over to the CEO and whispered that he should have worn a papal ring for the bankers to kiss as they came through the door. Laughing, Westphal shot back that "in a few months, I'll probably be on my knees begging them to keep the IPO on track."

He wasn't exactly joking. In fact, six weeks later he was walking past a Fidelity Investments office in downtown Boston and noticed that every stock price displayed on a big screen in its window was shown in red. The market had suddenly gone south with a vengeance, and the Dow Jones Industrial Average had just plunged more than five hundred points, its biggest drop since the September 11 terror attacks. One of the possibilities that kept him awake at night seemed about to become real, for trying to sell early-stage biotech shares in the wake of a market crash would be like panhandling during a hurricane. But when he talked to the J.P. Morgan bankers leading the IPO later that day, they made reassuring sounds. The downturn was no more than the expected stumbling of an aging bull, they said, and Sirtris's story was so special and hot that the market's vicissitudes likely wouldn't have much effect on demand for its shares.

Indeed, investment bankers were pressing Sirtris to seek at least $60 million from the IPO, which at first was more than Westphal thought prudent. He worried that its shares would wind up being sold at a demoralizing markdown if the expected demand didn't materialize. But he became significantly less worried after he was unexpectedly dealt a couple of wild cards in his high-stakes game with Wall Street: Two famous investors came forward out of the blue and jammed a small fortune into Sirtris's back pocket.

In early 2007, hedge fund tycoon John Henry, principal owner of the Boston Red Sox, and legendary former Fidelity fund manager Peter Lynch joined forces to lead a $36 million investment in Sirtris— Henry initiated the financing after getting excited about the potential of sirtuin-based drugs to retard aging. To Westphal the extraordinary infusion of cash couldn't have come at a better time, for it sent a clear message to the institutional investors who would mainly buy the biotech's IPO stock: Sirtris isn't desperate for your money, so forget trying to beat down its per-share price.

Just before the price was finally set in early May at a quite respectable $10 a share, netting $62.4 million, Westphal and his top lieutenants at Sirtris went on "the road show," a standard part of the IPO process, during which they crisscrossed the country to give presentations to potential investors. They felt like overbooked rock stars. At one point, Bohlin gleefully e-mailed Sirtris colleagues from his Black-Berry that they were "fortunate not to be riding with us right now on our way to the airport. Your CEO just did a complete change of clothes in the back seat of our SUV. Fortunately, Michelle [Dipp] is in the other car."

· · ·

With about $100 million added to its coffers by the IPO and the prior financing round in early 2007, Sirtris had little need to forge an alliance with a large drug company in the near term. But to maintain its momentum, Westphal needed to keep its stock price out of the doldrums. If the price sank much below what the shares had sold for at the IPO, and stayed there for an extended period, Sirtris would likely get a reputation as just another biotech that had failed to deliver—totally unjustified, perhaps, but that's life on Wall Street. Thus, he had to veer near enough to the vortex to generate positive media coverage, but not so close as to tarnish Sirtris's image.

His role as the "the great levitator" (his post-IPO nickname at Sirtris) didn't require as much prestidigitation as it did for many biotech CEOs. Research on sirtuins was taking off, and labs around the world were making advances that found their way into Wall Street analysts' reports on Sirtris, bolstering perceptions of it as a hot stock. Many of the findings sprang from labs overseen by Sirtris advisers. Indeed, Westphal had fashioned the biotech's far-flung group of advisers into a kind of melting pot for sirtuin research.

One of the scientific advisory board's latecomers was MIT's Guarente, the father of the sirtuin niche and cofounder of Elixir Pharmaceuticals. Elixir and Sirtris, recall, had long vied for recognition as the leading biotech based on aging research. A few months after the IPO, Sirtris announced that it had licensed exclusive rights to applications of an important discovery in Guarente's lab that suggested SIRT1 activators might reduce artery clogging. The announcement left little doubt that Sirtris had pushed ahead of Elixir in the sirtuin niche, and a few weeks later, Guarente was named cochairman of Sirtris's scientific advisory board.

But it wasn't all smooth sailing following the IPO. For one thing, Sirtris was having trouble getting FDA clearance to begin U.S. diabetes trials with SRT501. In keeping with a long-standing rule, the agency required the biotech to submit toxicity data on SRT501 from two quite different animal species before testing the drug in people. Typically, mice and dogs are used in such studies. All had gone well in the tests with rodents. But the canine part hadn't panned out—for some reason, SRT501 tends to pass through dogs' digestive tracts without being absorbed. (Curiously, that doesn't happen in rodents and people.) And despite the fact that scores of diabetes patients had already safely taken SRT501 during Sirtris's preliminary trial in India, the FDA insisted that human testing in the United States couldn't proceed until more animal data were available. One day when I visited Elliott after he'd had a particularly frustrating phone conversation with an agency official, I had a premonition that the eminently throwable sheep figurines on a shelf in his office weren't long for this world. But I was wrong—the mild-mannered clinical director never hurled a single one at his wall.

In September 2007, Wall Street analysts downgraded Sirtris's stock—the shares had simply gotten too pricey, they said. But doubters about the company had another thing coming from the great levitator. A few weeks earlier he'd decided to go ahead with the report on the research advances at Sirtris, and *Nature* had quickly accepted the paper for publication. (After leading the highly successful IPO, Westphal had more power to call the shots at the company.) As expected, Sirtris was required to divulge the structure of the compounds whose effects were highlighted in the paper. Among them was the first molecule that the company's researchers had identified as having real promise as a prototype drug. Dubbed SRT1720, it appeared to be a potent SIRT1 activator with salubrious effects in lab mice prone to diabetes.

Ironically, after Westphal had made the decision to reveal Sirtris's results with SRT1720, the compound began losing luster as a drug candidate. Mysteriously, it was found to cause nausea in monkeys. Luckily, the side effect didn't appear to be a general downside of SIRT1 activation, and quite different molecules soon replaced SRT1720 in the company's queue of development candidates. But by the end of 2007, SRT1720 was headed to the shelf. Thus, when the *Nature* paper appeared in late November, it divulged little of value to Sirtris's competition. Right after the paper was published, Sirtris's share price briefly spiked from the midteens to more than twenty dollars, establishing a high-water mark that would figure into the company's value when it was acquired a few months later.

The report in *Nature* opened a new chapter in the sirtuin story, helping to move it beyond the haze surrounding resveratrol. It described in detail how SRT1720 potently activates the SIRT1 enzyme in the test tube, and induces CR-like effects in rodents, including the lowering of insulin in different mouse models of diabetes and the forming of new mitochondria in muscles. Importantly, the rodents' weight was unchanged on the drug, so the effects couldn't be dismissed as a case of drug-induced appetite suppression inadvertently causing CR.

France's Auwerx chimed in a year later with a second report on SRT1720 showing that, like resveratrol, it boosts exercise endurance in mice and enhances the use of fat as fuel in their muscles. Hefty doses of SRT1720 even prevented weight gain on rich diets. The study, coauthored by Sirtris scientists, did show some differences between the actions of resveratrol and the novel SIRT1 activator. But it generally supported Sinclair's and Guarente's view that SIRT1 activation mimics calorie restriction. (It should be noted that not all of the mouse data on SRT1720 have been encouraging—a group at Pfizer reported in early 2010 that the compound failed to show benefits in diabetes-prone mice.)

The *Nature* paper burnished Sirtris's reputation as more than a hot stock—the company was beginning to look like a truly substantial drug developer. That impression was further strengthened by growing evidence that multiple members of mammals' sirtuin family of genes have major roles in metabolism. The family is defined by its seven members' resemblance to the SIR2 gene in yeast, which MIT's Guarente had famously identified as regulating life span. While the sirtuins have very complex functions that will take years to pin down, it seems that at least several of them join with SIRT1 to help slow aging when starvation looms. SIRT3, SIRT4, and SIRT5, for instance, appear to be involved in fuel switching during CR, boosting energy production from molecules other than sugar when it's in short supply. A 2006 study overseen by Harvard researcher Fred Alt, a Sirtris adviser, showed that SIRT6 helps keep DNA molecules pristine— when SIRT6 is disabled in mice, they show signs of premature aging and die soon after birth.

The rapid pace of discovery on the various sirtuins confronted Sirtris with a tricky decision: whether to continue betting the farm on SIRT1-based drugs or instead to devote a big portion of its budget to early-stage research on other members of the sirtuin family. Some of the biotech's top scientists pushed the latter as a way to mitigate risk—

if activating SIRT1 turned out to have major side effects, for example, the company would have fallbacks. Others favored maintaining the heavy emphasis on SIRT1, including an expansion of clinical testing with SRT501 to cancer patients. Westphal steered a middle course, supporting initial forays into the wider sirtuin space but not spreading bets so much that the ambitious SIRT1 program had to be scaled back. As chief levitator, he regarded the ability of clinical trials with SRT501 to generate investor interest as critical for maintaining Sirtris's momentum. Besides, he still felt that if SRT501 fulfilled his and Sinclair's dreams for it—a long shot, he readily conceded—it would change everything, and not just for Sirtris.

But would the powers that be at GlaxoSmithKline buy into the dream?

• • •

Talks between Glaxo and Sirtris had begun more than two years earlier when Tachi Yamada, then the British drug company's head of research and development, had explored a possible partnership with the biotech. It was obvious that Glaxo saw SIRT1 activators as potential successors to Avandia, a blockbuster for diabetes that it had launched in 1999. Avandia helped usher in a new class of medicines, called glitazones, that mitigate insulin resistance. But the glitazones can cause weight gain and other side effects that SIRT1-based drugs might dispense with while delivering similar benefits.

Glaxo wasn't interested in developing resveratrol (or SRT501) because of the murk surrounding its mechanism of action and the inability to patent it. And when its team arrived at Sirtris in early 2006, the biotech had little to show besides preliminary mouse data with resveratrol. "It was like, 'Who cares about that?'" Westphal told me later. "But they told us that if we got good results [with novel SIRT1 activators], they'd be very interested."

For the next two years Sirtris saw Glaxo as its leading competitor. But the rivalry turned to romance after Sirtris reported the promising results with SRT1720 in *Nature*. Glaxo R & D chief Moncef Slaoui, a dapper, cosmopolitan Moroccan immunologist who had succeeded

Yamada, wasted no time letting Sirtris know that the drug company's interest in a partnership was on the rise. In late 2007, he and Glaxo's head of drug discovery, Patrick Vallance, flew to Boston to meet Westphal and talk about combining forces on sirtuins.

In January 2008, Sirtris stirred further excitement by announcing at a major biotech conference that SRT501 had shown modest but statistically significant signs of efficacy in its early-stage trial with diabetes patients. After that, recalled Westphal, "the dance was fully on" with both Glaxo and several other prospective partners.

Sirtris's negotiations with Glaxo, however, were complicated by an ongoing CEO transition at the pharma company. In October 2007, it had announced that Andrew Witty would succeed Jean-Pierre Garnier as chief executive. As the changeover drew near, Glaxo seemed to lose interest in Sirtris. Toward the end of March 2008, Westphal recalled, "we hadn't heard from them for a month. I was really annoyed," and he decided to skip an upcoming research conference in New York that Glaxo had invited him to. "But Garen [Bohlin, Sirtris's chief operating officer] said, 'You have to go.' So I said, 'Okay, I'll do the dinner the night before'" the meeting and then leave.

After the March 31 dinner at Per Se, one of New York's few Michelin three-star restaurants, Slaoui took Westphal aside for a private chat. The impatient biotech CEO decided to take a fish-or-cut-bait stance, and to keep it simple he surprised Slaoui by proposing that Glaxo buy Sirtris outright, cutting short the protracted negotiations that would be necessary to work out who would control what in a partnership. A few days later Glaxo tentatively agreed to buy Sirtris for $22.50 a share, or $720 million.

When the deal was announced on April 22, many industry watchers seemed mystified. Underscoring the bafflement, Forbes called the deal "an expensive purchase of an essentially unproven medicine," adding that Sirtris had been trading for only about twelve dollars a share before the announcement.

Glaxo's logic wasn't all that hard to figure out, though. A year earlier, the New England Journal of Medicine had published a high-profile study linking Avandia to increased risk of heart attacks. The data on the issue weren't clear-cut, and some studies showed no appreciable signs that the Glaxo drug poses cardiac risk. But as the controversy about the drug escalated, the FDA told Glaxo to add a "black box" warning related to heart risks to its label. Then an FDA advisory panel recommended that in the future the agency should more rigorously scrutinize diabetes drugs' safety risks before approving them. Avandia's sales plunged, and some consumer advocates even called for it to be pulled off the market. If Sirtris's drugs worked, they would be just what the doctor ordered for what was ailing Glaxo in the diabetes market.

That wasn't all. Like other industry players, Glaxo has been beleaguered by the soaring costs of drug R & D. Industry-wide, the average cost of developing a drug in 2004 had risen to more than $860 million, probably too much to be sustained for long, according to a 2006 article in the *New England Journal of Medicine*. Worse, drug regulators are increasingly reluctant to approve, and insurers to pay for, new medicines that offer only incremental improvements over older, cheaper generic ones. The possibility that Sirtris's experimental drugs were CR mimetics—and thus able to counter many diseases of aging in a novel way with few side effects—promised some potent relief from this larger, long-standing malaise. As Witty commented after the acquisition was completed, "If we are right then [Sirtris] is a platform for multiple drugs. This is a very binary type of investment because if we are wrong we might get nothing, but this is exactly the type of thing that a company like GSK should have in its portfolio."

Glaxo also coveted Sirtris because Slaoui and Witty were planning to restructure their company's drug-discovery operation to infuse it with biotechlike speed and creativity. The plan called for the formation of small "discovery performance units," or DPUs, of ten to eighty people that would be more narrowly focused than Glaxo's traditional discovery groups—a DPU might work on a single drug target. Slaoui saw Sirtris as a ready-made, bellwether DPU, and because of that, Westphal felt Glaxo would take special pains not to crush Sirtris's entrepreneurial spirit.

Despite the deal's made-in-heaven look, Westphal was asked to make a major concession during the negotiations. Eager to keep him on staff after the acquisition, Glaxo insisted that he forgo his right to exercise all of his Sirtris stock options when the biotech was sold—in essence, it wanted to hold back a fourth of the pay to which he was contractually entitled upon a change in control at Sirtris and dole it out it to him over a four-year period, giving him an incentive to stay. That was fine with Sirtris's board, which didn't want Westphal to rock the boat with Glaxo, potentially putting the deal at risk. Indeed, Westphal informed me later, one board member "told me, 'Sorry, Christoph, you're just going to have to be thrown under the bus on this one.'"

After some tense interactions with his fellow directors, who weren't asked to take a similar hit to their postacquisition proceeds, Westphal agreed to the concession with the proviso that none of Sirtris's staffers would get less than he did in terms of stock-option vesting. What the heck, he'd wanted to stick around anyway, he told me after the deal was announced, and it often occurred to him that helping Sirtris realize its promise might well be the most important thing he ever did.

· · ·

When I dropped by Sirtris a year after the acquisition, Westphal seemed happy with the way things had gone. Keeping to plan, the mother ship, as he called Glaxo, had maintained its distance, largely leaving Sirtris to its own devices. Much of the biotech's staff had stayed, although there were a number of new names on its organization chart. The highest-up one was George Vlasuk, a former Wyeth executive who had been appointed president, a new position. While Westphal was still overseeing Sirtris, he was spending much of his time in a new role—he'd been tapped to lead a Glaxo unit that develops alliances with outside biotechs. Dipp, after working with Bohlin to integrate Sirtris with Glaxo, had also been drafted to help run the deal-making unit.

For his part, Sinclair was still serving as a scientific adviser to Sirtris, spearheading sirtuin studies at Harvard, and occasionally stirring controversy. In late 2008 he'd quit as a scientific adviser to Shaklee Corporation after a *Wall Street Journal* article suggested he was helping to promote a resveratrol-containing tonic Shaklee touted as "the world's best anti-aging supplement." In media stories about the episode, he expressed dismay about being portrayed on the Web as a kind of celebrity endorser of the product.

There had been some culture clashing. Westphal, for instance, had annoyed some visitors from the mother ship by declining to take part in the kind of all-day meetings they were used to. But the pluses, he said, had dwarfed the problems. "If we hadn't been acquired, our stock would be worth a lot less now," he told me—biotech shares had been crushed by the market crash. Like many other small biotechs, Sirtris would probably have had to curtail R&D to avoid a cash crunch. Instead, it was moving at full speed and had launched clinical tests with a novel SIRT1 activator to treat diabetes and inflammatory disorders such as psoriasis, initiated two cancer trials with SRT501, and, in concert with Glaxo and outside scientists, begun investigating its medicines' potential to treat many other diseases, from Alzheimer's to atherosclerosis.

While talking about the present and future with Westphal, I found myself also looking back, musing about how he and colleagues had managed to prove that anti-aging research can spawn a roaring commercial success.

Sinclair's sass, brass, and ambition were obvious contributors, as was Westphal's gift for getting a passel of big cats to act like Iditarod huskies pulling across Alaska. Luck—lucky timing, in particular—was also a factor. It's arguable that Elixir Pharmaceuticals came along a little too early to fully capitalize on its founders' exciting basic research on gerontogenes. As a result, it wound up licensing its leading drug candidate from another company—a relatively low-risk, low-reward strategy compared with Sirtris's big bet on SIRT1 activators. Just after Sirtris agreed to be acquired for a flabbergasting sum, Elixir withdrew a planned IPO for lack of investor interest.

But the chief reason for Sirtris's success, in my view, was that its underlying scientific premise, while possibly off in some details, was basically sound—at this point the balance of evidence indicates that resveratrol really does emulate key aspects of CR in animals, and that sirtuins at least partly mediate those effects. That explains why one study after another, including ones by researchers with no ties to

Sirtris, have appeared since the early 2000s suggesting that targeting SIRT1 and other sirtuins may confer a strikingly broad array of health benefits that would otherwise be hard to account for. The Sirtris team deserves credit for deftly riding this scientific wave, but the company wouldn't have gone far unless the wave had been truly powerful. So here's what I see as the moral of Sirtris's story: Sinclair/Westphal may be the Lennon/McCartney of twenty-first-century biotech, but you don't necessarily need another blue-moon pair like them to realize the promise of recent progress in anti-aging science. You just need to make a beeline, as they so singlemindedly did, for the Great Free Lunch.

12

THE GEORGE BURNS
SCENARIO

JUST A FEW weeks before his death in 1996 at age 100, George Burns was still visibly enjoying life, cracking wise at a Christmas party thrown by Frank Sinatra. France's Jeanne Calment, who holds the record for longevity, was similarly droll and unsinkable. She took up fencing at 85, was still riding a bicycle at 100, and at 121, a little more than a year before her death, released a CD, *Time's Mistress,* in which she reminisced to rap and other music. When a reporter at an annual party in her honor departed with the words "Until next year, perhaps?" she shot back, "I don't see why not! You don't look so bad to me."

Very old people with such élan are obviously rare. But I suspect that many who retain mental clarity in late life make their way toward something like Burns's and Calment's radiant rapprochement with old age. Multiple studies show that self-reported happiness among older people in reasonably good health is generally higher than among groups at younger ages. I don't want to sugarcoat old age—it isn't for sissies, as they say. But I'd love to see more well-tempered sages like Burns in the world, setting an example. Call it the George Burns scenario. This book, of course, is about the quest to make the scenario possible.

Critics of this quest, however, see little chance that anti-aging drugs would abet the scenario. They argue that the ability to brake aging may well engender a disastrous surfeit of needy oldsters gripped by greed and ennui. And these modern-day apologists have gotten quite heated up in recent years by signs that anti-aging breakthroughs aren't far off.

One of the most ardent apologist commentaries of recent years was penned by an undertaker, Thomas Lynch of Milford, Michigan, who urged readers of his 1999 *New York Times* op-ed piece (including, presumably, his potential future customers) to say "thanks, but no thanks" to "medicos pushing the big enchilada" of life extension, lest "like drunks with drink" we all get addicted to the pursuit of longer lives. Lynch, a whimsically sardonic writer, argued that anti-aging drugs would merely compound the damage done by the twentieth century's life-expectancy gains: "With all this extra time to kill, we went to war more, divorced more, aborted more and Kevorked [sic] more than any of the short-lived generations before us. . . . If our great-grandchildren will have to wait till 90 for the hard-won sense that came to us at 45, if they must endure double the incremental damages of serial monogamy, the 24-hour news cycle, easy-listening music, Tae Bo and infomercials, telemarketers and television preachers, no amount of Ritalin or Prozac or Viagra will make them well, however fit they seem to be at 150."

Two years later, Leon Kass, a University of Chicago professor who was soon to be named chairman of the Bush administration's Council on Bioethics, weighed in with a kind of apologist manifesto for the twenty-first century in *First Things,* a journal on religion and culture. "Let us resist the siren song of the conquest of aging and death," he wrote, for "the finitude of human life is a blessing for every human individual, whether he knows it or not. . . . The desire to prolong youthfulness is not only a childish desire to eat one's life and keep it; it is also an expression of a childish and narcissistic wish incompatible with devotion to posterity."

In 2003, the bioethics council mirrored Kass's views in a major report titled *Beyond Therapy: Biotechnology and the Pursuit of Happiness.* Anti-aging drugs, the council opined, might make people feel less urgency to get important things done before their time is up, making for a "life of lesser engagements and weakened commitments." They may make us more preoccupied with our own health, wealth, and pleasure, as well as "far less welcoming of children" and the sacrifices they require. Society might become top-heavy with oldsters weighed down by "disappointed hopes and broken dreams, accumulated mistakes and misfortunes . . . diminished ambition, insensitivity, fatigue, and cynicism." Frustrated younger people would "see before them only layers of their elders blocking the path." And perhaps worst of all, the drugs might lead to the nightmare of prolonged late-life morbidity and misery—what I call the Tithonus scenario.

The council further reflected that "one might wonder whether we have already gone too far in increasing longevity," and pondered whether it might be good to "roll back at least some of the increases made in the average human lifespan over the past century." After deliberating carefully, however, it decided not to call for banning things like antibiotics, hypertension drugs, and seat belts. Still, the decision seemed a rather close call.

Don't get me wrong—I share some of the apologists' concerns. Tampering with the aging process in an unprecedented way clearly calls for a close examination of the possible consequences, and the apologists deserve credit for urging us to be thoughtful. But I think a number of their arguments cry out for summary defenestration.

First, I see no reason to grapple with their favorite straw man, the coming of immortality. As I see it, immortality and "research" related to it are, shall we say, a bit too futuristic to matter to those of us of a certain age. (Such as our great-grandchildren.)

It's always dangerous to predict that a breakthrough like "engineered negligible senescence" won't happen for a very long time— the term, which refers to a reasonable facsimile of immortality, was coined by Aubrey de Grey, the clever, provocative editor-in-chief of the journal *Rejuvenation Research*. But it's just as perilous to freely extrapolate from today's technology to the biomedical equivalent of intergalactic travel. Consider how confident many people were in the 1960s that by now computers would be so intelligent they could pass as boxed humans (Remember *2001*'s HAL?), and that manned flights to other planets would be routine. While it may be possible to engineer something like negligible senescence in the distant future, I think it has a vanishingly small chance of happening while anyone now alive still is.* Thus, I regard apologist hand-wringing about the forthcoming "conquest of aging and death," to borrow Kass's phrasing, as akin to getting in a lather about the future outsourcing of American jobs to a race of robots in the Andromeda Galaxy.

The anti-aging drugs likely to matter in the near future will be CR mimetics that, with luck, will be capable of postponing the onset of major diseases of aging by five to ten years and perhaps of extending maximum life span by a comparable amount. Such drugs could launch a medical revolution as profound as the one engendered by the advent of vaccines and antibiotics. But don't count on them curing killer diseases or reversing heavy damage caused by decades of doing the wrong things—they'll mainly be preventatives, not rejuvenators. Let us not forget Eubie Blake's quip: "If I'd known I was gonna live this long, I'd have taken better care of myself."

Second, while most of the social and spiritual aspects of aging are beyond this book's scope, I find it impossible to quietly pass over the apologists' assertion that extending life would only increase anomie. Sounding off on this, Kass wrote: "If the human life span were increased even by only twenty years, would the pleasures of life increase proportionately? Would professional tennis players really enjoy playing 25 percent more games of tennis? Would the Don Juans of our world feel better for having seduced 1,250 women rather than 1,000? Having experienced the joys and tribulations of raising a family until the last had left for college, how many parents would like to extend the experience by another ten years?"

To me, such sentiments mainly call to mind the words of British

---

*As the abysmally low success rate in drug development shows, it's incredibly difficult to make even relatively minor adjustments in the operation of living systems. Thus, I feel that engineering the far more profound, complex changes needed for negligible senescence—even if we knew precisely what changes were required, which we don't—is about as likely to happen during, say, the next fifty years as the bioengineering of a little Dutch boy with hundreds of arms and thousands of fingers. Still, I admire de Grey's spirited, detail-oriented optimism, and I find his speculations about how we might someday arrest bodily decay, detailed in his 2007 book *Ending Aging*, quite intriguing—no other far-out anti-aging dreamer has as many interesting things to say as he does.

bioethicist John Harris ("Only the terminally boring are in danger of being terminally bored" by an extended life) and of Henry David Thoreau ("None are so old as those who have outlived enthusiasm").

Third, I think the apologists' characterization of the quest for anti-aging drugs as a form of me-generation selfishness is nonsense. There's nothing "childish and narcissistic" about wanting to remain in good health as long as possible, to minimize the elder-care burdens one places on one's spouse or kids (the most immediate representatives of posterity), to position oneself to give back for as long as possible, and to remain lucid as one comes to grips with mortality—even to rage against the dying of the light, if that feels right. Anti-aging drugs promise to abet all these things along with, yes, some prolonging of youthfulness.

. . .

Modern apologists often rail about the resource depletion and environmental destruction that presumably will occur if life expectancy is boosted, worsening population pressures. But let's say "thanks, but no thanks" to these Malthusian gloomsayers. For one thing, rising health and life spans have generally gone hand in hand in recent decades with falling birthrates and smaller families, and so it's not clear that a modest increase in life expectancy would really increase world population—it might even help reduce it over time. Further, lifestyle trumps raw population numbers—if everyone on earth burned as much fossil fuel as the typical American, for example, the planet would soon be toast, and that fact alone totally overshadows the risk to sustainability posed by a modest increase in population that may or may not occur. Besides, birthrates across the world are falling—even in developing countries such as Mexico and China they are now at levels that are expected to cause their populations to begin dropping within a few decades. Amazingly, China's labor supply is expected to be shrinking by as early as 2020.

Some apologists assert that it's unethical to devote millions of dollars to the pursuit of anti-aging drugs when so many people in the developing world still lack basic necessities. To be consistent, they also would have had to oppose the development of antibiotics, cholesterol-lowering drugs, and virtually every other life-extending therapy of the past century. Enough with the phony trade-offs.

Apologists fret that anti-aging drugs would enable the Stalins of the world to extend their reigns of terror. But should we halt work on all medical advances that may extend tyrants' lives? And why focus only on tyrants? Anti-aging drugs also promise to extend the lives of the world's Thomas Jeffersons, Jane Austens, Albert Einsteins.

What about the oncoming tsunami of "greedy geezers," as the New Republic heavy-handedly put it? Aged baby boomers supposedly will dominate America in coming years, gobbling ever larger chunks of the federal budget for entitlement programs, blocking local spending on schools and other programs that benefit the young, and running up monster deficits that will hobble future generations. To geezer-phobes, anti-aging drugs seem the very last thing we need.

Greedy-geezer alarmists have been issuing red alerts about the graying of America since the late 1970s. Their views were sardonically reflected in Christopher Buckley's 2007 novel Boomsday, whose heroine starts a national movement to avert economic ruin by giving tax breaks to boomers who kill themselves. (Shades of Trollope's novel The Fixed Period, in which the aged were supposed to be euthanized by law soon after retirement.) But the alarmists aren't black humorists. They're quite serious. Consider, for example, this incendiary assertion from authors William Strauss and Neil Howe, specialists on generational conflict: "Seniors suck the marrow from our bones through Social Security."

If you cut through the myths and spin that surround population aging, you can find issues that truly deserve careful consideration. (We'll get to some of them shortly.) But it takes a lot of cutting. While you're at it, keep in mind that greedy-geezer alarmists are allied to lobbyists for insurance companies and other business interests that would benefit from the rolling back or privatizing of federal entitlement programs for the elderly.

The idea that hidebound seniors will soon get a stranglehold on the body politic is one of the first things to cut. There's no gerontocratic bloc—older voters are as politically heterogeneous as the rest of the population. In U.S. presidential elections between 1980 and 2004, the voting patterns of citizens over 60 were nearly identical to those of voters between 30 and 59. This isn't surprising. As economist James Schulz and gerontologist Robert Binstock note in their 2006 book, *Aging Nation*, "old age is only one of many personal characteristics of aged people, and only one with which they may identify themselves." Boomers aren't about to change that as they reluctantly truck on toward the great Woodstock in the sky. While surveys suggest that the flower-power generation has gotten somewhat more politically conservative over time, more boomers voted for Barack Obama than for John McCain in the 2008 presidential election.

Claims that boosting life span will be economically ruinous also represent spin. In an eye-opening 2006 paper, University of Chicago economists Kevin Murphy and Robert Topel calculated that the years of life Americans gained between 1970 and the 2000, largely from postponing deaths caused by diseases of aging, were worth $61 *trillion* after subtracting the costs of health expenditures during the period. Their analysis was based on the commonsense idea that we're willing to spend large sums to lower our risk of death because life is quite precious even when its supply goes up. You can estimate just how dear it is by investigating, for instance, how much extra pay a person would demand to accept a one in ten thousand chance of getting killed on the job over the next year, a standard technique in economics on which Murphy and Topel's analysis was based. (In effect, economists have shown that William Blake inadvertently expressed a great socioeconomic truth when he declared that "life delights in life.")

Here's another thing: Per capita incomes across the globe have long risen in tandem with life expectancy. Research by economists David Bloom of Harvard and David Canning of Queen's University in Belfast suggests that this isn't a coincidence—longer life boosts productivity and national wealth. Long-living people tend to be healthier all their lives, and healthier workers are physically and mentally more robust, losing fewer workdays from illness. They're motivated to invest more in education and developing skills, because they expect to reap the benefits of such investments over longer periods. They save more for retirement, amassing capital to fuel investments and economic growth. Based on studies of several countries, Bloom and Canning estimated that a five-year rise in life expectancy in a nation boosts its annual per capita income growth by 0.3 to 0.5 percent—an impressive amount given that such growth has typically hovered around 2 percent.

Evolutionary theory also offers provocative insights that bear on the greedy-geezer debate. The insights have sprung in part from efforts to explain a deep conundrum: Why do women undergo menopause, rapidly losing the power of reproduction when they're still quite vibrant? Recall, the force is with us, from an evolutionary perspective, only while we can hurl our genes into the future. After we stop reproducing, natural selection's body-preserving influence fades and we crumble under the dual onslaught of deleterious, late-acting genes and random molecular damage. But women typically live for decades after menopause, and some for nearly half their lives. That makes menopause seem, against all odds, programmed by evolution.

In his seminal 1957 paper on the evolution of aging, George Williams proposed that menopause evolved to enable mothers to devote more resources to their existing children. If women indefinitely continued reproducing in the bad old days, he reasoned, they might well have wound up with too many mouths to feed or died in childbirth, which could have been fatal for their existing offspring in light of our species' extraordinarily long period of early-life neediness. Notice that Williams's theory highlights the important role in evolution of the transfer of resources from older to younger individuals, an idea pregnant with possibilities.

A problem with Williams's idea is that average life span before the dawn of civilization was probably about twenty years, making it difficult to see how evolution could have shaped a pattern that emerges much later in life than that. Still, there's evidence that even in the Stone Age a number of individuals may have lived much longer than the average. And a kindred idea to Williams's menopause theory, called the "grandmother hypothesis," suggests another reason menopause might have evolved. It holds that in highly social species like ours, older females' ability to assist in child care enables younger female relatives, including their daughters, to bear more children than they otherwise would. In other words, menopause is all about ensuring an adequate supply of wise, vigorous babysitters in groups with lots of blood ties and shared genes.

Support for the grandmother hypothesis has come from studies on highly social marine mammals whose older females are known to act as midwives and babysitters for younger ones' offspring. Female pilot whales, for example, stop reproducing at about age forty, yet often live into their sixties. Postreproductive female whales even continue lactating, enabling them to nurse other mothers' calves. Patterns of aging in social insects also suggest that specialists in nuturing can evolve striking longevity. "Nurse bees," for example, care for the young in honeybee colonies and can live many times longer than their sisters who forage outside the colony; it seems the nurses have revved-up stress responses that slow aging, enabling some of them to survive through winter so that they can jump-start the nurturing of new workers in the spring.

In 2003, Ronald Lee, a demographer at the University of California at Berkeley, added heft and precision to such concepts in a landmark paper delineating the key role that intergenerational transfers—not just material basics such as food and warmth but also leading and teaching—play in shaping patterns of aging in social species. While couched in technical terms, Lee's theory deserves much more attention than it has gotten. By highlighting the ancient importance of elders' ability to acquire and confer resources in ways that boost youngsters' chances of thriving, it offers an antidote to the divisive rhetoric of the geezers-versus-youth crowd.

· · ·

Like all profoundly life-altering technologies, anti-aging drugs will confront us with some taxing changes and difficult choices. The most pressing ones, in my view, concern health-care and retirement costs, which may be substantially greater than currently projected if many people extend their lives with the drugs in coming years. But I don't see economic disaster ahead. And anti-aging drugs may help keep some the costs of rising longevity in check even while boosting others.

Forecasters who argue that population aging will wreak economic havoc tend to fixate on a statistic called the "aged dependency ratio," which compares the number of people age sixty-five and over with the number of working-age people between twenty and sixty-four. Regardless of whether anti-aging drugs arrive in the near future, the aging of baby boomers will increase this ratio in coming years; in 2040, according to federal projections, there will be only 2.0 workers per Social Security recipient, compared with 3.3 in 2005.

But our society has already proved itself perfectly capable of thriving with a heavier load of dependents than we'll see as boomers retire. In fact, the total dependency ratio (which factors in both children and retirees) in the United States will never get as high in coming decades as it did when boomers were young dependents, according to federal projections. And as kids, boomers were nearly totally dependent on their parents; as retirees, many will have savings and part-time jobs. Indeed, economists who have closely examined the costs of providing for dependents in different age groups project that the total "support burden" in the United States will actually be smaller between 2030 and 2050 than it was between 1950 and 1970.

Still, we'll need to make some adjustments to the Social Security system as the 70-million-plus members of the hell-no-we-won't-go generation set out on their long good-bye—the oldest boomers turn sixty-five in 2011. The system's trust fund will be solvent until 2037, according to federal projections issued in early 2009. Forced retirements during the "Great Recession" that began in 2008 are likely drawing down the trust fund faster than expected, though. Moreover, tax revenues are projected to begin falling below program costs in 2016. Thus, it makes sense to rejigger the system soon in order to avoid forced, radical changes later.

The needed adjustments won't necessarily be wrenching if we act soon, according to Robert Butler, an all-around authority on aging and the founding director of the National Institute on Aging. In his 2008 book, *The Longevity Revolution,* he lays out a mix of modest changes that would maintain solvency, such as raising the limit on annual wages subject to Social Security taxes to $150,000 from $97,000 (as of 2007).

Strengthening the Social Security safety net, however, won't solve the twin problems of boomers' tendency to underestimate how long they're likely to live and the accompanying failure to save enough to avoid old-age penury—deficiencies that anti-aging drugs may exacerbate. It's probable that many will need to work past typical retirement age in order to ensure adequate incomes during their elongated later lives. This isn't news to most of us, nor does it represent cosmic injustice. Since 1950 the average retirement age in the United States has dropped by about six years, from sixty-eight to sixty-two, yet life expectancy has risen by about ten years.

A trend toward working later is already under way. After declining between 1950 and 1990, labor force participation rates of people sixty-five and over began to rise and recently reached 17 percent, up from 12 percent a decade earlier. Achieving the higher percentages that will probably be needed, though, will require a society-wide rethink on older workers, beginning with many employers' view that hiring them is a losing proposition. Since 1967, federal laws against age discrimination have helped address such prejudices. But it seems many employers regard the penalties that occasionally result from shunning older workers as just another cost of doing business.

It isn't hard to refute the standard bad rap on older workers, that they're unproductive. While they may not be as mentally quick and physically vigorous as younger employees, they offer compensatory pluses. They tend to have more knowledge and experience, less turnover and absenteeism, a stronger work ethic, and greater motivation and engagement than younger workers do.

Surveys show that boomers overwhelmingly expect to work in retirement—most would prefer part-time jobs—and that a major incentive to do so for about 90 percent of them is the desire to stay mentally and physically active. It seems that a lot of people agree with a piece of wisdom that George Burns expressed late in his life: "Age to me means nothing. I can't get old, I'm working. I was old when I was twenty-one and out of work. As long as you're working, you stay young."

But it's more difficult to counter what is probably the main thing that deters employers from hiring older workers: their risk of health problems and relatively high health-care costs. Indeed, one survey found that a large percentage of older persons wanted to work but more than half cited poor health as a reason for not doing so. A fourth of early retirees between sixty-two and sixty-four are too frail to do ordinary work, according to a 2003 estimate by the National Academy of Social Insurance. In my view, such statistics represent a clarion call to make well-tested CR mimetics available as quickly as possible.

More such clarion calls have been unintentionally issued by forecasters at the federal Centers for Medicare and Medicaid Services, or CMS. As the prevalence of chronic diseases rises among aging boomers, U.S. health-care spending will nearly double during the next decade, according to CMS economists. By 2017 it will account for 19.5 percent of the U.S. gross domestic product, or $4.3 trillion, up from 16.3 percent in 2003. Meanwhile, spending on Medicare, the federal health-care program for people over sixty-five, is projected to soar to $884 billion, from $427 billion in 2007.

We need to put a halter on runaway U.S. medical costs. But it behooves us to keep in mind that spending hefty sums on health isn't necessarily a bad thing. As economist Robert Fogel has written, "The increasing share of global income spent on health-care expenditures is not a calamity; it is a sign of the remarkable economic and social progress of our age." Besides, the health-care industry is a perennially vibrant part of the U.S. economy. Many jobs helping to provide care to the old and ailing aren't particularly high-paid or prestigious, but at least they can't be off-shored.

Health-care spending is an investment, not just a cost, adds Duke University's Kenneth Manton, for it reduces a nation's future health bills and at the same time boosts its long-term economic growth. Healthier workers can stay in the workforce later in life, preserving experienced "human capital" that would otherwise be lost, and they tend to have less disability and lower health-care costs later in life. Largely due to these factors, Medicare spending between 1994 and 2004 was actually $93 billion lower than federal forecasters had projected at the beginning of the period.

It's also salient that the widening use of costly new medical technologies, often in ways that do little to improve health, probably has more to do with the explosion in health-care spending than rising life expectancy does. A national health-care system that rationalizes the use of expensive procedures and cuts needless administrative costs would go a long way toward containing the explosion. In Japan, which has a national health-care system (the great majority of rich nations do, by the way), the percentage of citizens over sixty-five rose by nearly a third between 1980 and 1990, yet the proportion of GDP spent on health care went up only 1.6 percent during that period. Japan's system isn't perfect—certain kinds of specialists are in short supply, for example. But the fact that it has the world's highest life expectancy makes it hard to argue that it has kept health-care costs in check via Draconian denial of care.

Earlier chapters highlighted evidence that CR mimetics would reduce infirmity in late life, helping to avert the Tithonus scenario. We've been doing pretty well even without anti-aging drugs—old-age disability rates in the United States, as noted earlier, have fallen steadily for decades.

Still, the prevalence of chronic illness at later ages isn't necessarily dropping. And the way we've been buying time probably isn't sustainable. One reason is that gains in health span and life expectancy over the past few decades have largely stemmed from improved management of cardiovascular diseases. But as heart disease mortality has been shoved to older ages, diseases of aging have come to the fore that are more complex and harder to mitigate than clogged arteries and high blood pressure—most prominently cancer and Alzheimer's disease. Since President Nixon declared the war on cancer in 1971, the National Cancer Institute has spent more than $100 billion on it. Yet the death rate for cancer, adjusted for the size and age of the population, declined only 5 percent between 1950 and 2005. Meanwhile, heart disease mortality fell by 64 percent.

Alzheimer's disease is proving even more recalcitrant than cancer. While U.S. deaths from heart disease, stroke, breast cancer, and prostate cancer fell between 2000 and 2004, those due to Alzheimer's disease rose by nearly a third. In fact, Alzheimer's prevalence is inexorably rising along with the average age of the population and is projected to more than triple by midcentury in the United States, to about 16 million people.

Such projections bring home a harsh reality: The older we've collectively gotten, the harder it has become to increase our healthy life span. And for all its success over the past half century, the traditional approach to battling the Reaper—treating one disease at a time, often after symptoms appear and irreversible damage has been done—appears to be turning into a very costly game of diminishing returns.

It's time to bring on some free lunch.

• • •

The graying of nations around the world isn't the only reason to develop and deploy CR mimetics. Such drugs are also sorely needed to help counter the tidal wave of disease coming at us due to rising obesity rates.

In the U.S., some two-thirds of adults are overweight or obese. Recently, the number of obese Americans (those with a body mass index over 30) actually surpassed the number who are merely overweight (BMI between 25 and 30). Health-care costs for obese adults are about 42 percent higher than for normal-weight individuals, and medical spending on obesity-related conditions reached an estimated $147 billion in 2008, about 10 percent of all medical spending. Being overweight has even been linked to a greatly elevated risk of dementia in old age.* If present trends continue, an astounding 86 percent of U.S. adults will be overweight or obese by 2030, and related medi-

---

*A few studies suggest that being overweight doesn't raise the risk of fatal diseases and may actually protect against them. But I think they're misleading. Importantly, they're based on subjects' BMI at single points in time, often late in life, rather than cumulative lifetime exposure to the deleterious effects of visceral fat—the kind that lards up the liver and other organs in potbellied persons—which is very likely what matters. And they may lowball overweight's health fallout due to inclusion of people with chronic undiagnosed diseases, especially smoking-related lung conditions that play out over many years. Such people often are thin yet die young, effectively making the thin side of the weight spectrum seem riskier, and the fat side seem less risky, for reasons that have nothing to do with weight.

cal spending will account for about 17 percent of all U.S. health-care costs.

The obesity epidemic isn't limited to the United States, or even to the developed world. In China, for example, more than a fourth of adults are overweight or obese, and that proportion may well double over the next two decades. Supersizing is pandemic.

The most dismaying aspect of the pandemic is its spread to younger ages. About a third of U.S. children are overweight, and more than a fourth are obese or morbidly obese. These numbers presage a catastrophic rise in young people of what used to be regarded as old-age diseases. Over the past two decades, the incidence of type 2 diabetes among adolescents has increased by a factor of more than 10. Obese children as young as eleven have been found to have atherosclerosis, putting them at elevated risk of stroke and heart attacks in early adulthood. Shockingly, researchers have found that a number of overweight children viewed as normal and healthy by their parents, and even by physicians, actually have enlarged hearts, a sign of incipient heart disease. Kids under ten have shown signs of the condition.

In 2005, demographer S. Jay Olshansky at the University of Illinois at Chicago and colleagues made a dismaying prediction in the *New England Journal of Medicine:* "The steady rise in life expectancy during the past two centuries may soon come to an end" because of the obesity epidemic.

The pandemic's causes are hotly debated, as is the optimal way to address it. In my view, this debate is largely academic—the problem is so urgent that we need to throw everything at it that we safely can, from eliminating soda machines in schools to funding research on better weight-loss strategies to developing new drugs for obesity-related diseases. CR mimetics could play a major role in heading off the pandemic's worst fallout. Much of obesity's bodily harm probably stems from the fact that it induces a proinflammatory state that bears many of the hallmarks of accelerated aging. CR damps down such inflammation, and CR mimetics should do the same. In fact, lowering inflammation seems to be one of resveratrol's most important CR-like effects.

. . .

What's up next in the anti-aging quest?

Vendors of dietary supplements containing resveratrol, as well as their loyal customers, would probably dismiss that as a silly question— the quest has already reached a successful conclusion in their view. Poppers of other pills touted as scientifically validated youth preserva- tives likely feel the same way. I wish them well in their enthusiastic pursuit of placebo effects.

That's not to say the CR-like effects linked to resveratrol, rapamy- cin, and some other compounds are illusory—lab animals don't expe- rience placebo effects. But as I explained in chapter 10, the data on these agents, while highly promising, haven't reached an actionable level for me personally. And though I'm willing to take such com- pounds before they've been rigorously vetted in the kind of clinical trials required for FDA approval of drugs (which I doubt will happen for many years, if ever), I'm reluctant to do so until enough clinical data are available to let me make a reasonably well-informed decision about optimal dosing.

In other words, I'd like to have about the same level of knowl- edge about purported CR mimetics that's obtained from phase 2 tri- als of drugs during the three-phase clinical testing required for FDA approval. (Phase 2 usually yields a pretty good handle on efficacy, safety, and dosing; larger, longer phase 3 trials are aimed at provid- ing decisive data.) Reaching the phase 2 point in the development of anti-aging drugs would require further animal life-span studies and research on biomarkers of aging, followed by clinical trials that would likely last at least several years. (The length of the trials would be determined largely by the number of subjects enrolled—roughly speaking, the more people in a trial, the faster it yields statistically significant data about things like gene-activity levels correlated with slowed aging.)

Researchers backed by private foundations and the National Insti- tute on Aging are already carrying out much of the preclinical ground- work. Scientists at the Barshop Institute in Texas, for example, have launched a promising effort to conduct life-span studies in enchanting little primates called marmosets. Marmosets are almost as closely related to humans as are rhesus monkeys (the primates used for major studies on CR), yet they live only six to fifteen years, making life-span studies with them far faster, cheaper, and easier than ones with rhesus monkeys, which live up to forty years. The small primates also are readily available and aren't threatened or endangered, as are many larger ones. Using biomarkers of aging, scientists working with marmosets might be able to rapidly assess the efficacy of anti-aging drugs in a way that would say a lot about their probable effects in humans—perhaps in mere months.

Meanwhile, the NIA's calorie restriction study—the CALERIE project mentioned in chapter 7—promises new insights on biomarkers related to slowed aging.

Such research is paving the way for applied gerontology to become a major medical discipline. But unfortunately the nascent field is getting all dressed up with nowhere to go—the combination of big bucks and the political will necessary to carry out major clinical tests of anti-aging drugs is sorely lacking.*

How to fill this gap? The pharmaceutical industry doesn't care about aging—it's not a disease. Nutraceutical makers have the will but not the way—their products' relatively small profit margins won't pay for large clinical trials. In the fullness of time, nonprofit research foundations might get the job done. In late 2009, Sirtris's Christoph Westphal and David Sinclair launched a "healthy lifespan" institute that might undertake a major study of resveratrol's ability to add quality time. (Similar data might be collected, Westphal told me, in "postmarketing" studies on new diabetes drugs—the FDA requires pharmaceutical companies to conduct such investigations to check for side effects that don't appear in clinical trials.)

---

*I'm aware of only one recent clinical trial of anti-aging effects: DSM Nutritional Products, a vitamin seller based in the Netherlands, has sponsored a small, exploratory clinical trial of resveratrol pills' ability to induce gene-activity changes in muscle like those caused by CR. Insulin sensitivity and other biomarkers affected by CR were also assessed. Conducted at Washington University School of Medicine in St. Louis, the three-month, placebo-controlled trial employed modest doses of resveratrol, 75 milligrams a day. The results haven't been disclosed at this writing.

But the only player in a position to move the anti-aging quest forward in a fast, focused way is the federal government. Underscoring its key role as an enabler of cutting-edge science, a group of gerontologists joined forces a few years ago to urge the launching of a federal push on anti-aging research that would be comparable to the 1960s Apollo program. Its mission, laid out in a 2006 article in *The Scientist* by Jay Olshansky, Richard Miller, Robert Butler, and Daniel Perry, executive director of the Alliance for Aging Research, would be to develop and widely deploy interventions that brake aging enough to delay all aging-related diseases by about seven years. The program to realize this "longevity dividend" would have a yearly budget of $3 billion.

That's a sizable figure—the NIA's total annual budget is now about $1 billion—but it would be less than 1 percent of annual Medicare spending. And funding the development of anti-aging drugs promises far greater returns than spending in any other field of medical research. In a landmark 2005 study, health economist Dana Goldman and colleagues at the RAND Corporation compared the costs and benefits of ten medical technologies, including anti-aging drugs, that may be widely adopted in coming years. Assuming that anti-aging drugs would add ten healthy years to life expectancy, the researchers calculated that the medicines would buy an extra life-year for $8,790 (in 1999 dollars)—a bang per buck that greatly exceeded that for the other technologies. A medicine for Alzheimer's disease that cut its prevalence by a third was projected to cost nearly ten times as much for each year of added life, more than eighty thousand dollars.

RAND's analysis suggested that anti-aging drugs would boost overall health-care spending more than would the other technologies, mainly because the drugs would add many more life-years than would one-disease-at-a-time palliatives. While apologists may see that as a reason to halt research on anti-aging drugs, I think most of us would be willing to shell out a lot more than $8,790 for an extra year—especially if we could count on having the good health and sleek, youthful look of old mammals on CR during our extra time.

Trying to mimic CR isn't the only promising approach to developing anti-aging drugs. (Though it seems likely to be the most fruitful in the near term.) Pursuing medicines that emulate gerontogenes' effects, which as we've seen overlap those of CR, might also yield breakthroughs. And as the University of Michigan's Richard Miller notes, "the things we can do in the laboratory to play around with longevity are trivial compared to what nature long ago figured out how to do. By caloric restriction, we can get about a 50% increase in longevity . . . But nature, even within a specific group of mammals, can do a lot better. The shortest-lived rodents differ by a factor of 10 from the longest-lived rodents. The shortest-lived primates and the longest-lived primates differ by a factor of about 10 or so. If we want to make any real progress in this area, I think we need to figure out how nature does it."

Sadly, comparative gerontology, the niche devoted to figuring out how nature does it in animals such as bats and naked mole-rats, has long been one of biomedicine's poor cousins. Indeed, it's arguable that most of the lines of research covered in this book are lamentably underfunded, given their growing potential to shed light on aging and how to slow it. Mindful of that, the advocates of the proposed longevity-dividend program recommend that about a third of its budget be devoted to basic studies on the biology of aging.

The payoffs of the program, they argue, "may very well be one of the most important gifts that our generation can give" to future ones. I could not agree more. The George Burns scenario is within our grasp, and we could instigate it while most people now alive are still around if we take account of what has happened in aging science and seize the day. The anti-aging revolution probably won't enable all of us to live as long as Burns did. But by helping us age as gracefully as he did, it should let an unprecedented number of us aspire to his timelessness. As Burns once quipped, "You can't help getting older, but you don't have to get old." Words to remember from a wise guy to the end.

ACKNOWLEDGMENTS

I BECAME FASCINATED with gerontology in the late 1990s when, as a medical writer at *Fortune* magazine, it struck me that aging is by far the leading risk factor for practically every disease I covered. Yet while we collectively devote great effort and huge sums of money to countering other risk factors—smoking, environmental carcinogens, high cholesterol—we generally treat aging as a given to be thoughtlessly accepted. That wouldn't have seemed odd a few years earlier. But like many journalists covering science and medicine, I was aware of the then-recent discovery of gene mutations that could dramatically extend life span in worms, flies, and even mice. Surely, I thought, those basic discoveries must be lighting the way toward interventions that could retard human aging, an advance whose importance would dwarf that of any other potential medical breakthrough. But exactly what way forward was being lit?

My first pass at answering that question led to a three-part series on gerontology and its commercial spinoffs that *Fortune* ran in 1999. This book is a direct descendant of that series, and so my first thanks go to editors at the magazine—Peter Petre, Tim Smith, and John Huey—who made the effort possible. Once hooked, I repeatedly circled back to the topic, culminating with a 2007 *Fortune* cover story on Sirtris Pharmaceuticals overseen by another talented editor, Jim Aley. I'm also indebted to the *Wall Street Journal*'s former Boston bureau chief, Gary Putka, who ushered into the paper two breaking news stories about resveratrol that I wrote in 2006. When it all came together as a book idea, the persistent optimism of my agent, Lisa Adams, was immensely valuable. Editors Tim Sullivan, now at Basic Books, and Brooke Carey of Current ably helped shape the book.

Of the numerous scientists who led me along the path gerontology is blazing, none was more important than Steve Austad at the University of Texas and Richard Miller at the University of Michigan. Both spent many hours explaining things, guiding me around land mines, and generously vetting a draft of this book—their input immeasurably improved it. (Whatever flaws have remained, of course, are my own.) Many others helped fill me in over the past decade, including but not limited to Johan Auwerx, Andrzej Bartke, Nir Barzilai, Joe Baur, Holly Brown-Borg, Rochelle Buffenstein, Robert Butler, Rafa de Cabo, Judith Campisi, Aubrey de Grey, John Denu, Susan Doctrow, Caleb Finch, Kevin Flurkey, Dana Goldman, Lenny Guarente, Denham Harman, David Harrison, Anthony Hulbert, Donald Ingram, Tom Johnson, Matt Kaeberlein, Brian Kennedy, Cynthia Kenyon, Michael Klass, Valter Longo, Bernard Malfroy, Kenneth Manton, Ed Masoro, Jay Olshansky, Tom Perls, Pere Puigserver, Eric Ravussin, Arlan Richardson, George Roth, Gary Ruvkun, Dave Sharp, Stephen Spindler, Randy Strong, Heidi Tissenbaum, James Vaupel, Huber Warner, and Richard Weindruch.

I owe special thanks to David Sinclair, and to his assistant Susan DeStefano, for enabling me to closely track the Sinclair lab's unfolding work on sirtuins over the past five years, and to Christoph Westphal, who welcomed me as a kind of embedded reporter at Sirtris during its early years and also spent countless hours keeping me abreast. Many others at Sirtris kindly let me sit in on their work lives, including Garen Bohlin, Paul Brannelly, Michelle Dipp, Peter Elliott, Mike Jirousek, Phil Lambert, Walter Lunsmann, Jill Milne, Karl Normington, Susannah Walpole, and Roger Xie. A number of Sirtris directors and advisers also lent aid, including Rich Aldrich, John Freund, Steve Hoffman, Thomas Maniatis, Paul Schimmel, and Phillip Sharp.

My most inspiring sources were three amazing centenarians, Catherine McCaig, and Loren and Augusta Reid, who talked with me at length about their lives and the nature of graceful aging. Their geniality, humor, and equanimity were, and will always be, my touchstones when thinking about aging.

Finally, my wife, Alicia Russell, provided critical initial enthusiasm and unwavering support for this project, and my kids, Quentin and Claire, supplied the best of all possible reasons for me to ponder what it would take to live as long as possible.

NOTES

PROLOGUE: THE MOUSE STUDY THAT ROARED

2 **published that day by the journal:** J. A. Baur et al., "Resveratrol improves health and survival of mice on a high-calorie diet," *Nature* 444 (2006): 337–42.

1: THE WILD OLD WICKED MYSTERY OF AGING

7 **The "strange second puberty":** See D. Wyndham, "Versemaking and Lovemaking—W. B. Yeats' 'Strange Second Puberty': Norman Haire and the Steinach Rejuvenation Operation," *Journal of the History of the Behavioral Sciences* 39 (2003): 25–50; and Brenda Maddox, *Yeats's Ghosts: The Secret Life of W. B. Yeats* (New York: HarperCollins, 1999), 277–81.

7 **he enthused to his publisher:** As cited in Ann Saddlemyer, *Becoming George: The Life of Mrs. W. B. Yeats* (Oxford: Oxford University Press, 2002), 476.

7 **he wrote fifty poems:** See Wyndham, "Versemaking and Lovemaking—W. B. Yeats' 'Strange Second Puberty': Norman Haire and the Steinach Rejuvenation Operation," *Journal of the History of the Behavioral Sciences.*

8 **elicited ridicule:** Ibid.

8 **Sigmund Freud:** See A. Kahn, "Regaining Lost Youth: The Controversial and Colorful Beginnings of Hormone Replacement Therapy in Aging," *Journals of Gerontology: Biological Sciences* 60A (2005): 142–47; and C. Sengoopta, "'Dr. Steinach coming to make old young!': sex glands, vasectomy and the quest for rejuvenation in the roaring twenties," *Endeavour* 27 (2003): 122–26.

8 **inspired stranger things:** Ibid.

10 **Peter Medawar:** Biographical material on Medawar was drawn from his autobiography, Peter Medawar, *Memoir of a Thinking Radish* (Oxford & New York: Oxford University Press, 1986), and a memoir by Medawar's wife, Jean Medawar, *A Very Decided Preference: Life with Peter Medawar* (New York & London: W. W. Norton & Co., 1990).

11 **Haldane's approval:** See B. Charlesworth, "John Maynard Smith: January 6, 1920–April 19, 2004," *Genetics* 168 (2004): 1105–9.

11 **gave pride of place**: Peter Medawar, *The Uniqueness of the Individual* (New York: Basic Books, 1957).

11 **the blind watchmaker**: Richard Dawkins, *The Blind Watchmaker: Why the Evidence of Evolution Reveals a Universe Without Design* (New York: W. W. Norton & Co., 1987).

12 **a certain morbid fascination**: See Robert Arking, *Biology of Aging: Observations and Principles*, 2nd ed. (Sunderland, MA: Sinauer Associates, 1998).

12 **cadre of brilliant evolutionary thinkers**: See Mark Ridley, *Evolution*, 3rd ed. (Oxford: Blackwell Publishing, 2004), 14–19.

12 **Alex Comfort**: See Claire Rayner, "Alex Comfort: Dazzling intellectual whose prolific output of novels, poetry and philosophy remains overshadowed by a sex manual," *Guardian*, March 28, 2000.

12 **Austad has suggested**: Steven N. Austad, *Why We Age: What Science Is Discovering about the Body's Journey through Life* (New York: John Wiley & Sons, 1997), 98.

14 **E. Ray Lankester**: E. Ray Lankester, *On Comparative Longevity in Man and the Lower Animals* (London: Macmillan & Co., 1870).

14 **350-pound, 19-foot pike**: Ibid., 57.

14 **negligible senescence**: See Caleb E. Finch, *Longevity, Senescence, and the Genome* (Chicago: University of Chicago Press, 1994), 134 and 206–36. See also John C. Guerin's Web site on negligible senescence at http://www.agelessanimals.org/.

14 **Ming the Clam**: Richard Alleyne, "Clam, 405, is oldest animal ever," *The Daily Telegraph*, October 31, 2007.

14 **bowhead whales**: James Meek, "The old man of the sea," *Guardian*, November 17, 2000. See also J. C. George et al., "Age and growth estimates of bowhead whales (Balaena mysticetus) via aspartic acid racemization," *Canadian Journal of Zoology* 77 (1999): 571–80.

15 **pine named Methuselah**: Finch, *Longevity, Senescence, and the Genome*, 229.

15 **George C. Williams has pointed out**: G. C. Williams, "Pleiotropy, natural selection, and the evolution of senescence," *Evolution* 11 (1957): 398–411.

15 **fact that Darwin himself**: See B. Charlesworth, "Fisher, Medawar, Hamilton and the Evolution of Aging," *Genetics* 156 (2000): 927–31.

15 **vestiges of their dead ideas**: See L. A. Gavrilov and N. S. Gavrilova, "Evolutionary Theories of Aging and Longevity," *The Scientific World* 2 (2002): 339–56.

15 **In a note he jotted**: As quoted in Finch, *Longevity, Senescence, and the Genome*, 669–70.

16 **August Weismann**: See "August Freidrich Leopold Weismann," *Encyclopedia of World Biography*, 2nd ed. (Farmington Hills, MI: Gale Group, 1997).

16 **Echoing Wallace**: August Weismann, "The Duration of Life," *Essays Upon Heredity and Kindred Biological Problems* (Oxford: Clarendon Press, 1891).

16 **young biologists of the day**: See Gavrilov and Gavrilova, "Evolutionary Theories of Aging and Longevity," *The Scientific World*.

16 **lost his early enthusiasm**: See Michael R. Rose, *Evolutionary Biology of Aging* (New York: Oxford University Press, 1991), 6–7.

16 **theorists on gerontology's fringe**: See, for example, J. T. Bowles, "Shattered: Medawar's test tubes and their enduring legacy of chaos," *Journal of Medical Hypotheses* 54 (2000): 326–39.

16 **highlighted the blunder**: Medawar, "Old Age and Natural Death," *The Uniqueness of the Individual*, 3.

17 **his alternative theory**: Medawar, "An Unsolved Problem of Biology," *The Uniqueness of the Individual*, 43–48.

18 **Robert Arking neatly put it**: Arking, *Biology of Aging: Observations and Principles*, 112.

18 **public talk at the**: See John Brockman, "George C. Williams: A Package of Information," *The Third Culture: Beyond the Scientific Revolution* (New York: Simon & Schuster, 1995).

18 **Echoing Weismann**: See R. G. Winther, "An obstacle to unification in biological social science: Formal and compositional styles of science," *Graduate Journal of Social Science* 2 (2005).

19 **a 1966 book**: George C. Williams, *Adaptation and Natural Selection: A Critique of Some Current Evolutionary Thought* (Princeton, NJ: Princeton University Press, 1966).

19 **a 1957 paper**: Williams, "Pleiotropy, natural selection, and the evolution of senescence," *Evolution*.

19 **The son of a banker**: Personal communication, Doris Williams.

20 **Emerson "was all in favor of death"**: See Brockman, "George C. Williams: A Package of Information," *The Third Culture: Beyond the Scientific Revolution*.

20 **Emerson was no lightweight**: E. O. Wilson and C. D. Michener, "Alfred Edwards Emerson," *Biographical Memoirs*, vol. 53 (Washington, DC: National Academies Press, 1982): 159–77.

20 **1871 book**: Charles Darwin, *The Descent of Man, and Selection in Relation to Sex* (London: John Murray, 1871).

20 **Quaker-inspired pacifism**: See R. Tobey, "Political Context of Scientific Thought," *Reviews in American History* 22 (1994): 277–82.

21 **essence of Williams's theory**: Williams, "Pleiotropy, natural selection, and the evolution of senescence," *Evolution*.

22 **"I predict that no human"**: Ibid.

23 **"this conclusion banishes"**: Ibid.

23 **February evening in 1977**: Biographical details on Kirkwood were drawn from Tom Kirkwood, *Time of Our Lives: The Science of Human Aging* (New York: Oxford University Press, 1999). See also T. B. L. Kirkwood, "Evolution of ageing," *Nature* 270 (1977): 301–4.

24 **Kirkwood and a colleague, Robin Holliday**: T. B. Kirkwood and R. Holliday, "The evolution of ageing and longevity," *Proceedings of the Royal Society B: Biological Sciences* 205 (1979): 531–46. See also T. B. L. Kirkwood and M. R. Rose, "Evolution of Senescence: late survival sacrificed for reproduction," *Philosophical Transactions of the Royal Society B: Biological Sciences* 332 (1991): 15–24.

25 **landmark study in the late 1970s:** See Rose, *The Long Tomorrow: How Advances in Evolutionary Biology Can Help Us Postpone Aging* (New York: Oxford University Press, 2005), 39–45.

25 **Writing about it later:** Ibid.

26 **"scientific porn":** See "Pieces of the Puzzle: An Interview with Leonid A. Gavrilov, Ph.D.," *Journal of Anti-Aging Medicine* 5 (2002): 255–63.

26 **Looking into the future:** Medawar, *Memoir of a Thinking Radish*, 197–98.

CHAPTER 2: RADICALS RISE UP

27 **drugs he'd invented:** S. Melov et. al., "Extension of life-span with superoxide dismutase/catalase mimetics," *Science* 289 (2000): 1567–69.

27 **trained in his native France:** Lewis D. Solomon, *The Quest for Human Longevity* (New Brunswick, NJ: Transaction Publishers, 2006): 61–71.

28 **we buy it:** See "300% Growth in Antioxidant Food and Drink Products in Five Years as Consumers Fight Age," *NewswireToday,* July 27, 2007.

29 **press release:** Press release jointly issued by Eukarion and the Buck Institute, "First Successful Use of Drugs to Extend Life Span," August 2000.

29 **media loved it:** See "Worm has turned in anti-aging fight," *Belfast Telegraph,* September 1, 2000; Mark Henderson, "Humble worm holds secret of eternal youth," *The Times* (London), September 1, 2000; Grace Mclean, "Worms could help us all to wriggle out of old age," *Daily Record* (Glasgow), September 1, 2000; Laura Johannes, "Worm Study May Aid Research on Aging," *Wall Street Journal,* Section B, September 1, 2000; Ellen Licking, "Longer Life—For Worms," *BusinessWeek,* September 11, 2000.

29 **Canadian radio talk show:** Personal communication, Bernard Malfroy.

30 **interview with the British newspaper:** James Meek, "New drug for longer life," *Guardian* (London), June 30, 2001: 1.

30 **nematodes get old:** See online article by B. M. Zuckerman and S. Himmelhock, "Nematodes as models to study aging," http://www.wormbase .org.

31 **Denham Harman:** Biographical details on Harman drawn from personal communication, Denham Harman; and from K. Kitani and G. O. Ivy, "I thought, thought, thought for four months in vain and suddenly the idea came—an interview with Denham and Helen Harman," *Biogerontology* 4 (2003): 401–12.

32 **a Russian scientist:** See "Bogomolets & the Longer Life," *Time,* June 17, 1946.

33 **free radical:** For more on free radicals, see Nick Lane, *Oxygen: The Molecule That Made the World* (Oxford: Oxford University Press, 2002).

34 **led by Rebeca Gerschman:** R. Gerschman et al., "Oxygen Poisoning and X-irradiation: A Mechanism in Common," *Science* 119 (1954): 623–26.

34 **Barry Commoner:** B. Commoner et al., "Free radicals in biological materials," *Nature* 174 (1954): 689–91.

35 **molecules as ephemeral:** See K. J. A. Davies and W. A. Pryor, "The evolution of Free Radical Biology & Medicine: A 20-year history," *Free Radical Biology & Medicine* 39 (2005): 1263–64.

35 **In 1956 Harman first presented**: D. Harman, "Aging: A theory based on free radical and radiation chemistry," *Journals of Gerontology* 11 (1956): 298–300.

36 **blue-green protein**: See I. Fridovich, "The trail to superoxide dismutase," *Protein Science* 7 (1998): 2688–90.

36 **"children with a new toy"**: Ibid.

36 **impossibly fast rate**: See Lane, *Oxygen: The MoleculeThat Made the World*, 201.

36 **it became clear**: See Fridovich, "The trail to superoxide dismutase," *Protein Science*.

36 **like tiny dragons, spit superoxide**: Ibid.

37 **feeding antioxidants**: See D. Harman, "Free radical theory of aging: dietary implications," *American Journal of Clinical Nutrition* 25 (1972): 839–43.

37 **Comfort reported**: A. Comfort et. al., "Effect of ethoxyquin on the longevity of C3H mice," *Nature* 229 (1971): 254–55.

37 **large daily doses**: Medawar, *Memoir of a Thinking Radish*, 201.

37 **data were mixed**: See Harman, "Free radical theory of aging: dietary implications," *American Journal of Clinical Nutrition*.

38 **Harman acknowledged**: Kitani and Ivy, "I thought, thought, thought for four months in vain and suddenly the idea came—an interview with Denham and Helen Harman," *Biogerontology*.

39 **may well age faster**: See S. N. Austad and D. M. Kristan, "Are mice calorically restricted in nature?" *Aging Cell* 2 (2003): 201–7.

39 **second big idea**: D. Harman, "The biologic clock: the mitochondria?" *Journal of the American Geriatrics Society* 20 (1972): 145–47; and Harman, "The aging process," *Proceedings of the National Academy of Sciences* (hereafter, *PNAS*) 78 (1981): 7124–28.

40 **theory was later honed**: See T. Ozawa, "Mitochondrial DNA Mutations and Age," in *Towards Prolongation of the Healthy Lifespan: Practical Approaches to Intervention*, ed. D. Harman, R. Holliday, and M. Meydani (New York: New York Academy of Sciences, 1998).

40 **Ozawa and colleagues**: M. K. Hayakawa et al., "Age-related extensive fragmentation of mitochondrial DNA into minicircles," *Biochemical and Biophysical Research Communications* 226 (1996): 369–77.

40 **Stadtman**: E. R. Stadtman, "Protein Oxidation and Aging," *Science* 257 (1991): 1220–24.

41 **Other biotechs were riding**: See Aeolus form S-1 filed with the SEC on June 7, 2007; Christopher Tritto, "Say Goodbye to Metaphore," *St. Louis Business Journal*, December 16, 2005; and http://juvenon.com/.

41 **idea didn't pan out**: B. J. Day, "Catalytic antioxidants: a radical approach to new therapeutics," *Drug Discovery Today* 9 (2004): 557–66.

41 **seed idea for Eukarion's**: Personal communication, Bernard Malfroy.

43 **Catch-22 problem**: Personal communication, Bernard Malfroy; and "Modex Announces Licensing Agreement with Eukarion for a Small Molecule to Treat Radiation-Induced Skin Damage," *Business Wire*, October 5, 2001.

43 **compounds failed**: A. C. Bayne and R. S. Sohal, "Effects of superoxide dismutase/catalase mimetics on life span and oxidative stress resistance in the housefly, Musca domestica," *Free Radical Biology & Medicine* 32 (2002): 1229–

34; and M. Keaney and D. Gems, "No increase in lifespan in Caenorhabditis elegans upon treatment with the superoxide dismutase mimetic EUK-8," *Free Radical Biology & Medicine* 34 (2003): 277–82.

43   **encouraging data**: S. R. Doctrow et al., "Salen manganese complexes: Multifunctional catalytic antioxidants protective in models for neurodegenerative diseases of aging," in *ACS Symposium Series* (Washington, DC: American Chemical Society, 2005): 319–47.

43   **2003 mouse study**: R. Liu et al., "Reversal of age-related learning deficits and brain oxidative stress in mice with superoxide dismutase/catalase mimetics," *PNAS* 100 (2003): 8526–31.

44   **early April 2002**: Personal communication, Bernard Malfroy.

44   *Boston Business Journal:* "Australian firm snaps up Eukarion," *Boston Business Journal,* December 17, 2004.

44   **deal also included**: Personal communication, Bernard Malfroy.

45   **drug candidate failed**: See Christopher Tritto, "Say Goodbye to Metaphore," *St. Louis Business Journal,* December 16, 2005; and "ActivBiotics' Intellectual Property Assets, Including Drug Product Candidates, Sold for $3.5 Million," *PRNewswire,* April 2, 2008.

45   **Aeolus**: Aeolus press release, 2007.

45   **controversy ignited**: See L. Pauling, "Evolution and the Need for Ascorbic Acid," *PNAS* 67 (1970): 1643–48; L. Pauling, "Are Recommended Daily Allowances for Vitamin C Adequate?" *PNAS* 71 (1974): 4442–46; and Lane, *Oxygen: The Molecule That Made the World,* 171–93.

47   **abet free radical damage**: See B. Halliwell, "Vitamin C: antioxidant or pro-oxidant in vivo?" *Free Radical Research* 25 (1996): 439–54.

47   **iron plus vitamin C**: See A. E. Fisher and D. P. Naughton, "Iron supplements: the quick fix with long-term consequences," *Nutrition Journal* 3 (2004): 2; and A. E. Fisher and D. P. Naughton, "Vitamin C contributes to inflammation via radical generating mechanisms: a cautionary note," *Medical Hypotheses* 61 (2003): 657–60.

47   **large doses of vitamin C**: M. Khassaf et al., "Effect of vitamin C supplements on antioxidant defence and stress proteins in human lymphocytes and skeletal muscle," *Journal of Physiology* 549 (part 2, 2003): 645–52.

47   **cited such interference**: G. Bjelakovic et al., "Mortality in Randomized Trials of Antioxidant Supplements for Primary and Secondary Prevention," *Journal of the American Medical Association* 297 (2007): 842–57.

48   **four major clinical studies**: See E. Guallar et al., "An Editorial Update: Annus horribilis for Vitamin E," *Annals of Internal Medicine* 143 (2005): 143–45.

49   **Big Idea, Version 3.0**: See J. H. Chen et al., "DNA damage, cellular senescence and organismal ageing: causal or correlative?" *Nucleic Acids Research* 35 (2007): 7417–28; K. C. Kregel and H. J. Zhang, "An integrated view of oxidative stress in aging: basic mechanisms, functional effects, and pathological considerations," *American Journal of Physiology, Regulatory, Integrative and Comparative Physiology* 292 (2007): R18–36; and Caleb E.

Finch, *The Biology of Human Longevity: Inflammation, Nutrition, and Aging in the Evolution of Lifespans* (Burlington, MA: Academic Press, 2007).

50 **picture is murkier:** See S. E. Schriner et al., "Extension of murine life span by overexpression of catalase targeted to mitochondria," *Science* 308 (2005): 1909–11; and Y. C. Jan et al., "Overexpression of Mn Superoxide Dismutase Does Not Increase Life Span in Mice," *Journals of Gerontology Series A*, July 24, 2009.

3: HAGRID'S BAT AND THE SABER-TOOTHED SAUSAGE

52 **finding made it:** A. J. Podlutsky et al., "A New Field Record for Bat Longevity," *Journals of Gerontology* 60A (2005): 1366–68.

52 **the longevity quotient:** S. N. Austad and K. E. Fischer, "Mammalian Aging, Metabolism, and Ecology: Evidence from the Bats and Marsupials," *Journals of Gerontology* 46 (1991): B47–53.

52 **fit the usual pattern:** See Austad, *Why We Age*, 85; and R. Buffenstein, "Negligible senescence in the longest living rodent, the naked mole-rat: insights from a successfully aging species," *Journal of Comparative Physiology B* 178 (2008): 439–45.

53 **more interesting explanation:** Williams, "Pleiotropy, natural selection, and the evolution of senescence," *Evolution*.

54 **133 species of bats:** See A. K. Brunet-Rossinni and S. N. Austad, "Aging studies on bats: a review," *Biogerontology* 5 (2004): 211–22.

54 **best predictor of long life span:** G. S. Wilkinson and J. M. South, "Life history, ecology and longevity in bats," *Aging Cell* 1 (2002): 124–31.

57 **tracking opossums:** S. N. Austad, "Retarded senescence in an insular population of Virgina opossums," *Journal of Zoology* 229 (1993): 695–708.

59 **theory also handily explained:** J. Miquel et al., "Effects of temperature on the life span, vitality and fine structure of Drosophila melanogaster," *Mechanisms of Ageing and Development* 5 (1976): 347–70.

59 **Max Rubner:** See Austad, *Why We Age*, 72–73.

59 **cells of a rhinoceros:** See A. J. Hulbert and P. L. Else, "Membranes and the setting of energy demand," *Journal of Experimental Biology* 208 (2005): 1593–99.

59 **Raymond Pearl:** See Austad, *Why We Age*, 76–78.

60 **honeybees were forced:** T. J. Wolf and P. Schmid-Hempel, "Extra loads and foraging lifespan in honeybee workers," *Journal of Animal Ecology* 58 (1989): 943–54.

60 **Austad and a colleague highlighted:** Austad and Fischer, "Mammalian Aging, Metabolism, and Ecology: Evidence from the Bats and Marsupials," *Journals of Gerontology*.

61 **In 1980, researchers reported:** J. M. Tolmasoff et al., "Superoxide dismutase: Correlation with life-span and specific metabolic rate in primate species," *PNAS* 77 (1980): 2777–81.

61 **study led by Rajindar Sohal:** R. S. Sohal et al., "Relationship between

antioxidant defenses and longevity in different mammalian species," *Mechanisms of Ageing and Development* 3 (1990): 217–27.

61   **research by Gustavo Barja**: M. Lopez-Torres et al., "Maximum life span in vertebrates: Relationship with liver antioxidant enzymes, glutathione system, ascobate, urate, sensitivity to peroxidation, true malondialdehyde, in vivo H2O2, and basal and maximum aerobic capacity," *Mechanisms of Ageing and Development* 70 (1993): 177–99.

62   **Checking it out**: H. H. Ku et al., "Relationship between mitochondrial superoxide and hydrogen peroxide production and longevity of mammalian species," *Free Radical Biology & Medicine* 15 (1993): 621–27; and G. Barja et al., "Low mitochondrial free radical production per unit O2 consumption can explain the simultaneous presence of longevity and high aerobic metabolic rate in birds," *Free Radical Research* 21 (1994): 317–27.

62   **little brown bats**: A. K. Brunet-Rossinni, "Reduced free radical production and extreme longevity in the little brown bat (Myotis lucifugus) versus two non-flying mammals," *Mechanisms of Ageing and Development* 125 (2004): 11–20.

62   **free-tailed bats**: Brunet-Rossinni and Austad, "Aging studies on bats: a review," *Biogerontology*.

62   **huge amounts of sugar**: D. J. Keegan, "Aspects of the assimilation of sugars by Rousettus aegyptiacus," *Comparative Biochemistry and Physiology Part A* 58 (1977): 349–52.

64   **Mole-rats were introduced**: See J. M. Jarvis and P. W. Sherman, "Heterocephalus glaber," *Mammalian Species* 706 (2002): 1–9.

64   **remained little-studied**: Personal communication, Rochelle Buffenstein.

64   **Alexander theorized**: Elizabeth Pennisi, "Not just another pretty face—naked mole rat," *Discover*, March 1986.

64   **went public with the idea**: J. U. M. Jarvis, "Eusociality in a mammal: cooperative breeding in naked mole-rat colonies," *Science* 212 (1981): 571–73.

64   **up to three hundred NMRs**: Jarvis and Sherman, "Heterocephalus glaber," *Mammalian Species*.

66   **"Milton the Mole-Rat"**: Personal communication, Rochelle Buffenstein.

66   **the senior consort**: R. Buffenstein and J. U. M. Jarvis, "The Naked Mole Rat—A New Record for the Oldest Living Rodent," *Science of Aging Knowledge Environment*, no. 21 (2002).

66   **NMRs basically don't age**: T. P. O'Connor et al., "Prolonged longevity in naked mole-rats: Age-related changes in metabolism, body composition and gastrointestinal function," *Comparative Biochemistry and Physiology Part A* 133 (2002): 835–42.

67   **no more likely to die**: R. Buffenstein, "Negligible senescence in the longest living rodent, the naked mole-rat: insights from a successfully aging species," *Journal of Comparative Physiology B* 178 (2008): 439–45.

67   **in 2008 she proposed**: Ibid.

67   **social-insect queens**: See J. D. Parker et al., "Decreased expression of Cu-Zn superoxide dismutase 1 in ants with extreme lifespan," *PNAS* 101 (2004): 3486–89.

68 **mystery deepened**: O'Connor et al., "Prolonged longevity in naked mole-rats: Age-related changes in metabolism, body composition and gastrointestinal function," *Comparative Biochemistry and Physiology.*

68 **glutathione peroxidase**: B. Andziak et al., "Antioxidants do not explain the disparate longevity between mice and the longest-living rodent, the naked mole-rat," *Mechanisms of Ageing and Development* 126 (2005): 1206–12.

68 **NMRs' heart mitochondria**: A. J. Lambert et al., "Low rates of hydrogen peroxide production by isolated heart mitochondria associate with long maximum lifespan in vertebrate homeotherms," *Aging Cell* 6 (2007): 607–18.

68 **riddled with free radical**: B. Andziak et al., "High oxidative damage levels in the longest-living rodent, the naked mole-rat," *Aging Cell* 5 (2006): 463–71.

69 **NMRs operate at**: O'Connor et al., "Prolonged longevity in naked mole-rats: Age-related changes in metabolism, body composition and gastrointestinal function," *Comparative Biochemistry and Physiology.*

69 **low fasting blood sugar**: R. Buffenstein and M. Pinto, "Endocrine function in naturally long-living small mammals," *Molecular and Cellular Endocrinology* (2008), in press.

69 **Egyptian fruit bats'**: See Brunet-Rossinni and Austad, "Aging studies on bats: a review," *Biogerontology.*

69 **vampire bats'**: M. B. Freitas et al., "Effects of short-term fasting on energy reserves of vampire bats (Desmodus rotundus)," *Comparative Biochemistry and Physiology B* 140 (2005): 59–62.

69 **Baltimore Longitudinal Study of Aging**: G. S. Roth et al., "Biomarkers of caloric restriction may predict longevity in humans," *Science* 297 (2002): 811.

69 **shoot up by 500 percent**: Personal communication, Rochelle Buffenstein.

69 **Bats can rev**: Brunet-Rossinni and Austad, "Aging studies on bats: a review," *Biogerontology.*

70 **doses of hydrogen peroxide**: N. Labinskyy et al., "Comparison of endothelial function, superoxide and hydrogen peroxide production, and vascular oxidative stress resistance between the longest-living rodent, the naked mole-rat, and mice," *American Journal of Physiology—Heart and Circulatory Physiology* 291 (2006): H2698–704.

70 **cells from little brown bats**: J. M. Harper et al., "Skin-derived fibroblasts from long-lived species are resistant to some, but not all, lethal stresses and to the mitochondrial inhibitor rotenone," *Aging Cell* 6 (2006): 1–13.

70 **analyzed lipids**: A. J. Hulbert et al., "Oxidation-resistant membrane phospholipids can explain longevity differences among the longest-living rodents and similar-sized mice," *Journals of Gerontology Series A* 61 (2006): 1009–18.

70 **"membrane pacemaker" hypothesis**: A. J. Hulbert, "Life, death and membrane bilayers," *Journal of Experimental Biology* 206 (2003): 2302–11.

71 **Chaudhuri's assay**: A. Pierce et al., "A novel approach for screening the proteome for changes in protein conformation," *Biochemistry* 45 (2006): 3077–85.

71 **proteins from NMRs' liver cells**: V. I. Perez et al., "Protein stability and

resistance to oxidative stress are determinants of longevity in the longest-living rodent, the naked mole-rat," *PNAS* 106 (2009): 3059–64.

71 **Proteins from long-lived bats:** A. B. Salmon et al., "The long lifespan of two bat species is correlated with resistance to protein oxidation and enhanced protein homeostasis," *Journal of the American Federation of Societies for Experimental Biology*, February 24, 2009.

71 **implanting a gene:** C. J. Cretekos et al., "Regulatory divergence modifies limb length between mammals," *Genes & Development* 22 (2008): 141–51.

4: THE GENES THAT COULDN'T BE

73 **great minds:** C. Cerf and V. Navasky, *The Experts Speak: The Definitive Compendium of Authoritative Misinformation* (New York: Pantheon Books, 1984).

74 **Highlighting this point:** S. J. Olshansky et al., "In search of Methuselah: estimating the upper limits to human longevity," *Science* 250 (1990): 634–40.

74 **One estimate: Rose,** *The Long Tomorrow: How Advances in Evolutionary Biology Can Help Us Postpone Aging,* 97.

74 **A more formidable number:** G. M. Martin, "Genetic syndromes in man with potential relevance to the pathobiology of aging," *Birth Defects: Original Articles Series 14* (1978): 5–39.

76 **history of nematode research:** Andrew Brown, *In the Beginning Was the Worm* (New York: Columbia University, 2003).

76 **fanciers have found it:** See D. R. Denver et al., "Phylogenetics in Caenorhabditis elegans: An Analysis of Divergence and Outcrossing," *Molecular Biology and Evolution* 20 (2003): 393–400.

76 **strolled into his garden:** See Howard Ferris's Web site at the University of California at Davis, http://plpnemweb.ucdavis.edu.

77 **telescoping life span:** See T. E. Johnson, "Subfield History: Caenorhabditis elegans as a System for Analysis of the Genetics of Aging," *Science of Aging Knowledge Environment*, August 28, 2002.

77 **appeared in** *Nature:* M. Klass and D. Hirsh, "Non-ageing developmental variant of Caenorhabditis elegans," *Nature* 260 (1976): 523–25.

78 **In 1983, he reported:** M. R. Klass, "A method for the isolation of longevity mutants in the nematode Caenorhabditis elegans and initial results," *Mechanisms of Ageing and Development* 22 (1983): 279–86.

79 **Johnson had gotten interested:** Personal communication, Tom Johnson.

81 **paper was accepted:** D. B. Friedman and T. E. Johnson, "A mutation in the age-1 gene in Caenorhabditis elegans lengthens life and reduces hermaphrodite fertility," *Genetics* 118 (1988): 75–86.

82 **study finally appeared:** T. E. Johnson, "Increased life-span of age-1 mutants in Caenorhabditis elegans and lower Gompertz rate of aging," *Science* 249 (1990): 4908–12.

82 **there was little excitement:** Personal communication, Gary Ruvkun.

83 **report boldly concluded:** C. Kenyon et al., "A C. elegans mutant that lives twice as long as wild type," *Nature* 366 (1993): 404–5.

83 *Scientific American Frontiers:* Alda's interview in 2000 with Kenyon is at http://www.pbs.org/saf/transcripts/transcript1003.htm.

84 **A bookish polymath:** Biographical details on Kenyon were drawn from personal communication; Alex Crevar, "As the Worm Turns," *Georgia Magazine*, 2004; and Stephen S. Hall, *Merchants of Immortality: Chasing the Dream of Human Life Extension* (Boston: Houghton Mifflin Company, 2003).

85 **They publicized the finding:** *Worm Breeder's Gazette*, vol. 12, no. 3, June 15, 1992.

85 **a gene called daf-23:** S. Gottlieb and G. Ruvkun, "Daf-2, daf-16 and daf-23: genetically interacting genes controlling Dauer formation in Caenorhabditis elegans," *Genetics* 137 (1994): 107–20.

85 **researcher James Thomas:** Personal communication, Gary Ruvkun.

86 **A tall, droll, avuncular man:** Biographical details on Gary Ruvkun drawn from personal communication, Gary Ruvkun; and Ingfei Chen, "The Drifter," *Science of Aging Knowledge Environment,* October 23, 2003.

87 **his lab sifted:** J. Z. Morris et al., "A phosphatidylinositol-2-OH kinase family member regulation longevity and diapause in Caenorhabditis elegans," *Nature* 382 (1996): 536–39.

87 **mom-wowing gerontogene discovery:** K. D. Kimura et al., "daf-2, an insulin receptor-like gene that regulates longevity and diapause in Caenorhabditis elegans," *Science* 277 (1997): 942–46.

88 **Japanese study:** H. Kim et. al., "Detection of mutations in the insulin receptor gene in patients with insulin resistance by analysis of single-stranded conformational polymorphisms," *Diabetologia* 35 (1992): 261–66.

88 **record would later show:** S. Ogg et al., "The Forkhead transcription factor DAF-16 transduces insulin-like metabolic and longevity signals in C. elegans," *Nature* 389 (1997): 994–99; and K. Lin et al., "daf-16: An HNF-2/forkhead family member that can function to double the life-span of Caenorhabditis elegans," *Science* 278 (1997): 1319–22.

89 **the Rome of aging genes:** See S. S. Lee et al., "DAF-16 target genes that control C. elegans life-span and metabolism," *Science* 300 (2003): 644–47.

89 **a gene, dubbed "methuselah":** Y. J. Lin et al., "Extended life-span and stress resistance in the Drosophila mutant Methuselah," *Science* 282 (1998): 943–46.

89 **Leonard Hayflick:** L. Hayflick, "'Anti-Aging' Is an Oxymoron," *Journals of Gerontology, Series A* 59 (2004): 573–78.

90 **one of the subtexts:** See L. Partridge and D. Gems, "Beyond the evolutionary theory of ageing, from functional genomics to evo-gero," *Trends in Ecology and Evolution* 21 (2006): 334–40.

90 **a 2000 study:** D. W. Walker et al., "Evolution of lifespan in C. elegans," *Nature* 405 (2000): 296–97.

91 **among the first to shed light:** J. R. Vanfleteren, "Oxidative stress and ageing in Caenorhabditis elegans," *Biochemistry Journal* 292 (1993): 605–8; and P. L.

Larsen, "Aging and resistance to oxidative damage in Caenorhabditis elegans," *PNAS* 90 (1993): 8905–9.

91 **resistant to heat stress:** G. J. Lithgow et al., "Thermotolerance of a long-lived mutant of Caenorhabditis elegans," *Journals of Gerontology: Biological Sciences* 49 (1994): B270–76.

92 **worms' smell and taste:** J. Apfeld and C. Kenyon, "Regulation of lifespan by sensory perception in Caenorhabditis elegans," *Nature* 402 (1999): 804–9.

92 **GenoPlex lasted:** Personal communication, Tom Johnson.

5: THE REALLY STRANGE THING ABOUT DWARFS

94 **a Buddhist priest:** Mitosi Tokuda, "An Eighteenth Century Japanese Guide-Book on Mouse-Breeding," *Journal of Heredity* 26 (1935): 481–84.

95 **rage for "fancy" mice:** For details on the history of lab mice, see Lee M. Silver, *Mouse Genetics: Concepts and Applications* (Oxford: Oxford University Press, 1995); Terri Peterson Smith, "Mouse Work," *Invention and Technology* 23 (Summer 2007); and Herbert C. Morse's historical perspective on the lab mouse at http://www.informatics.jax.org.

96 **The Snell dwarf:** G. D. Snell, "Dwarf, A New Mendelian Recessive Character of the House Mouse," *PNAS* 15 (1929): 733–34.

96 **obituary in the *New York Times*:** "Obituary; Gen. Tom Thumb," *New York Times*, July 18, 1883, 8.

97 **In 1972, a study:** N. Fabris et al., "Lymphocytes, Hormones and Ageing," *Nature* 240 (1972): 557–59.

97 **The first blow:** R. Silberberg, "Articular Aging and osteoarthrosis in dwarf mice," *Pathology & Microbiology (Basel)* 38 (1972): 417–30.

97 **The following year:** J. G. M. Shire, "Growth Hormone and Premature Ageing," *Nature* 245 (1973): 215–16.

98 **In 1976, Gary Schneider:** G. B. Schneider, "Immunological competence in Snell-Bagg pituitary dwarf mice: response to the contact-sensitizing agent oxazolone," *American Journal of Anatomy* 145 (1976): 371–93.

98 **published in a now-defunct:** Personal communication, Andrzej Bartke.

98 **Studies in rats:** See A. V. Everitt et al., "The effects of hypophysectomy and continuous food restriction, begun at ages 70 and 400 days, on collagen aging, proteinuria, incidence of pathology and longevity in the male rat," *Mechanisms of Ageing and Development* 12 (1980): 161–72.

99 **chapter of a 1990 book:** D. E. Harrison, *Genetic Effects on Aging II* (Caldwell, NJ: The Telford Press, 1990): 435–56.

100 **Bartke began investigating:** Personal communication, Andrzej Bartke.

100 **abnormally rapid brain aging:** R. W. Steger et al., "Premature ageing in transgenic mice expressing different growth hormone genes," *Journal of Reproduction and Fertility* (supp.) 46 (1993): 61–75.

100 **German scientists reported:** R. Wanke et al., "The GH transgenic mouse as an experimental model for growth research: clinical and pathological studies," *Hormone Research* 3 (1992): 74–87.

101 **A native of Poland**: Biographical details on Bartke drawn from Ingfei Chen, "The Mouse That Roared," *Science of Aging Knowledge Environment*, 2003; and personal communication, Andrzej Bartke.

102 **Schaible had glanced down**: Personal communication, Robert Schaible.

103 **study published by the *New England Journal***: D. Rudman et al., "Effects of Human Growth Hormone in Men Over 60 Years Old," *New England Journal of Medicine* 323 (1990): 1–6.

103 **postdocs in Bartke's lab**: Personal communication, Holly Brown-Borg.

104 **a spectacular 50 percent longer**: H. M. Brown-Borg et al., "Dwarf mice and the ageing process," *Nature* 384 (1996): 33.

104 **Bartke's heart sank**: Personal communication, Andrzej Bartke.

105 **Flurkey's new study**: Personal communication, Kevin Flurkey.

106 **Prop-1**: M. W. Sornson et al., "Pituitary lineage determination by the Prophet of Pit-1 homeodomain factor defective in Ames dwarfism," *Nature* 384 (1996): 327–33.

107 **stunted, mutant mice**: K. T. Coschigano et al., "Assessment of growth parameters and life span of GHR/BP gene-disrupted mice," *Endocrinology* 141 (2000): 2608–13; K. Flurkey et al., "Lifespan extension and delayed immune and collagen aging in mutant mice with defects in growth hormone production," *PNAS* 98 (2001): 6736–41.

107 **age with amazing grace**: B. A. Kinney et al., "Evidence that age-induced decline in memory retention is delayed in growth hormone resistant GH-R-KO (Laron) mice," *Physiology and Behavior* 72 (2001): 653–60; and B. A. Kinney et al., "Evidence that Ames dwarf mice age differently from their normal siblings in behavioral and learning and memory parameters," *Hormones and Behavior* 39 (2001): 277–84.

107 **defenses are unusually robust**: H. M. Brown-Borg, "Hormonal regulation of longevity in mammals," *Ageing Research Reviews* 6 (2007): 28–45.

107 **fad came under fire**: See "Are Claims for Growth Hormone Bulked Up?" *Harvard Health Letter* 24 (April 1999): 1–3; and J. Takala et al., "Increased Mortality Associated with Growth Hormone Treatment in Critically Ill Adults," *New England Journal of Medicine* 341 (1999): 785–92.

107 **best obituary**: See Rich Miller's home page at http://www-personal.umich.edu.

108 **jointly investigated patterns**: I. Dozmorov et al., "Array-based expression analysis of mouse liver genes: effect of age and of the longevity mutant Prop1df," *Journals of Gerontology Series A* 56 (2001): B72–80.

108 **much-cited essay**: R. A. Miller, "Kleemeier award lecture: are there genes for aging?," *Journals of Geronotology Series A* 54A (1999): B297–307.

109 **played a leading role**: See R. A. Miller et al., "Big mice die young: early life body weight predicts longevity in genetically heterogeneous mice," *Aging Cell* 1 (2002): 22–29.

109 **led by Norman Wolf**: B. J. Deeb and N. S. Wolf, "Studying Longevity and morbidity in giant and small breeds of dogs," *Veterinary Medicine* (supp.) 89 (1994): 702–13.

110 **dwarf fruit flies**: M. Tatar et al., "A mutant Drosophila insulin receptor homolog that extends life-span and impairs neuroendocrine function," *Science*

292 (2001): 107–10; and D. J. Clancy et al., "Extension of life-span by loss of CHICO, a Drosophila insulin receptor substrate protein," *Science* 292 (2001): 104–6.

110 **professional baseball players:** See T. T. Samaras and L. H. Storms, "Impact of height and weight on life span," *Bulletin of the World Health Organization* 70 (1992): 259–67.

110 **prone to various cancers:** See G. D. Batty et al., "Adult height in relation to mortality from 14 cancer sites in men in London (UK): evidence from the original Whitehall study," *Annals of Oncology* 17 (2006): 157–66.

110 **shorter men:** See F. Kee et al., "Short Stature and Heart Disease: Nature or Nurture?" *International Journal of Epidemiology* 26 (1997): 748–56.

110 **Laron syndrome:** Z. Laron, "Laron Syndrome (Primary Growth Hormone Resistance or Insensitivity): The Personal Experience 1958–2003," *Journal of Clinical Endocrinology & Metabolism* 89 (2004): 1031–44.

110 **"little people of Krk":** V. Saftic et al., "Mendelian Diseases and Conditions in Croatian Island Populations: Historic Records and New Insights," *Croatian Medical Journal* 47 (2006): 543–52.

110 **Munchkins:** See http://www.kansasoz.com/munchkins; and Dan Barry, "No Ordinary Coroner, No Ordinary Life," *New York Times*, February 18, 2007.

111 **animals with low growth hormone:** See H. M. Brown-Borg, "Hormonal regulation of longevity in animals," *Ageing Research Reviews*, 2007.

112 **lived 26 percent longer:** M. Holzenberger et al., "IGF-1 receptor regulates lifespan and resistance to oxidative stress in mice," *Nature* 421 (2003): 182–87.

112 **mice with low brain levels:** L. Kapeler et al., "Brain IGF-1 Receptors Control Mammalian Growth and Lifespan through a Neuroendocrine Mechanism," *PLoS Biology* 6 (2008).

112 **deleterious effects on learning and memory:** See W. E. Sonntag et al., "Growth hormone and IGF-I modulate local cerebral glucose utilization and ATP levels in a model of adult-onset growth hormone deficiency," *American Journal of Physiology—Endocrinology and Metabolism.* 291 (2006): E604–10.

112 **extended the life spans of mice by 18 percent:** M. Bluher et al., "Extended longevity in mice lacking the insulin receptor in adipose tissue," *Science* 299 (2003): 572–74.

113 **quashed hopes:** See Robert Arking, *Biology of Aging*, 332–33.

113 **Mouse 11C:** Personal communication, Andrzej Bartke; and "Time runs out for world's oldest mouse," Southern Illinois University press release, January 15, 2003.

6: THAT OLD MAGIC

115 **Loren Reid:** Biographical details on Loren Reid drawn from personal communication; and Loren Reid, *Hurry Home Wednesday: Growing Up in a Small Missouri Town, 1905–1921* (Columbia, MO: University of Missouri Press, 1978).

117 **famous longevity illusionist:** See Austad, *Why We Age: What Science Is*

Discovering about the Body's Journey through Life, 1–2; and Westminster Abbey on Parr at http://www.westminster-abbey.org.

118 **bitingly skeptical tome:** William J. Thoms, *Human Longevity: Its Facts and Fictions* (London: John Murray, Albermarle Street, 1873).

118 **ran a cover story:** See B. J. Willcox et al., "Secrets of Healthy Aging and Longevity from Exceptional Survivors around the Globe: Lessons from Octogenarians to Supercentenarians," *Journals of Gerontology Series A* 63 (2008): 1181–85; and A. Leaf, "Every day is a gift when you are over 100," *National Geographic,* January 1973: 92–119.

118 **rigorous studies on centenarians:** See Willcox et al., "Secrets of Healthy Aging and Longevity from Exceptional Survivors around the Globe: Lessons from Octogenarians to Supercentenarians," *Journals of Gerontology Series A.*

118 **number of centenarians:** Thomas J. Perls, Margery Hutter Silver, with John F. Lauerman, *Living to 100: Lessons in Living to Your Maximum Potential at Any Age* (New York: Basic Books, 1999), 10.

118 **According to the Census Bureau:** See http://www.census.gov; and Carl Bialik, "Living to 100 May Be Easier Than Counting Those Who've Made It," *Wall Street Journal,* April 11, 2008.

119 **Death rates of old people:** R. Rau et al., "Continued Reductions in Mortality at Advanced Ages," *Population and Development Review* 34 (2008): 747–68.

119 **Between 1950 and 2002:** Ibid.

119 **reductions in childhood infections:** See C. E. Finch and E. M. Crimmins, "Inflammatory exposure and historical changes in human life-spans," *Science* 305 (2004): 1736–39.

119 **before the U.S. Senate:** Vaupel's testimony can be found at http://sageke .sciencemag.org.

119 **one British bookmaker:** Miles Brignall and Patrick Collinson, "William Hill slashes odds on living to 100," *Guardian,* April 28, 2007.

119 **British writer Ronald Blythe:** Ronald Blythe, *The View in Winter: Reflections on Old Age* (New York: Harcourt Brace Jovanovich, 1979).

120 **Aristotle portrayed:** *The Oxford Book of Aging,* ed. Thomas R. Cole and Mary G. Winkler (Oxford: Oxford University Press, 1994): 23–26.

120 **historian Gerald Gruman:** G. J. Gruman, "A History of Ideas about the Prolongation of Life: The Evolution of Prolongevity Hypothesis to 1800," *Transactions of the American Philosophical Society* 56 (1966): 1–102.

120 **the Roman poet Lucretius:** Ibid.

120 **Roman statesman Cicero:** Ibid.

121 **black humor:** *The Oxford Book of Aging,* 68.

121 **Francis Bacon:** Ibid., 34–35.

121 **Schopenhauer:** Ibid., 41.

121 **physician George Miller Beard:** Thomas R. Cole, *The Journey of Life: A Cultural History of Aging in America* (Cambridge, UK: Cambridge University Press, 1992): 163–67.

121 **novelist Anthony Trollope:** Ibid., 168–70.

122 **physician William Osler:** Ibid., 170–74.

122 **revisionist leader on aging:** See Gruman, "A History of Ideas About the

Prolongation of Life: The Evolution of Prolongevity Hypothesis to 1800," *Transactions of the American Philosophical Society*; Luigi Cornaro, *Discourses on a Sober and Temperate Life: Translated from the Italian Original* (London: Benjamin White, 1768); and the biographical sketch of Cornaro at http://www.boglewood.com/cornaro/xb26.html.

124 **Benjamin Franklin speculated:** See Gruman, "A History of Ideas About the Prolongation of Life: The Evolution of Prolongevity Hypothesis to 1800," *Transactions of the American Philosophical Society*.

124 **Marquis de Condorcet:** Ibid.

124 **economist's seminal work:** Ibid.

125 **apologists and their opponents:** See L. Kass, "L'Chaim and Its Limits: Why Not Immortality," *First Things* 113 (2001): 17–24.

125 **modern Malthusians:** See James H. Schulz and Robert H. Binstock, *Aging Nation: The Economics and Politics of Growing Older in America* (Baltimore: Johns Hopkins University Press, 2006).

125 **remarkable statistics:** J. Evert et al., "Morbidity profiles of centenarians: survivors, delayers, and escapers," *Journals of Gerontology Series A* 58 (2003): 232–37.

125 **In another study:** S. L. Andersen et al., "Cancer in the oldest old," *Mechanisms of Ageing and Development* 126 (2005): 263–67.

125 **escaped major diseases:** Perls, Silver, and Lauerman, *Living to 100,* 109.

125 **don't need to enter assisted-living:** D. C. Willcox et al., "Life at the extreme limit: phenotypic characteristics of supercentenarians in Okinawa," *Journals of Gerontology Series A* 63 (2008): 1201–8.

126 **The upshot, says Perls:** Perls, Silver, and Lauerman, *Living to 100,* 18.

126 **nearly two-thirds of 34:** M. H. Silver et al., "Cognitive functional status of age-confirmed centenarians in a population-based study," *Journals of Gerontology Series B* 56 (2001): 134–40.

126 **A Danish team reported:** K. Andersen-Ranberg et al., "Healthy centenarians do not exist, but autonomous centenarians do: a population-based study of morbidity among Danish centenarians," *Journal of the American Geriatrics Society* 49 (2001): 900–908.

126 **Baltimore Longitudinal Study:** See http://www.grc.nia.nih.gov.

126 **"successful aging movement":** See J. W. Rowe and R. L. Kahn, "Human aging: usual and successful," *Science* 237 (1987): 143–49.

127 **Satchel Paige:** Quoted in Schulz and Binstock, *Aging Nation,* 180.

127 **study on Danish twins:** A. M. Herskind et al., "The heritability of human longevity: a population-based study of 2872 Danish twin pairs born 1870–1900," *Human Genetics* 97 (1996): 319–23.

127 **Nir Barzilai:** Personal communication, Nir Barzilai.

127 **New England study's subjects: Perls,** Silver, and Lauerman, *Living to 100,* 59.

128 **none is obese:** Ibid., 71.

128 **Scanning local newspapers:** Perls, Silver, and Lauerman, *Living to 100,* 130.

128 **Some gerontologists dismissed:** Ibid.

129 **telltale analysis:** T. T. Perls et al., "Life-long sustained mortality advantage of siblings of centenarians," *PNAS* 99 (2002): 8442–47.

129 **Two years later, Barzilai:** G. Altzmon et al., "Clinical phenotype of families with longevity," *Journal of the American Geriatrics Society* 52 (2004): 274–77.

129 **first one was identified:** F. Schachter et al., "Genetic associations with human longevity at the APOE and ACE loci," *Nature Genetics* 6 (1994): 29–32.

130 **Perls developed a passion:** See Henry Chesbrough, "Centagenetix: Building a Business Model for Genetic Longevity," Harvard Business School case study, 2001.

130 **won a $150,000 grant:** Ibid.

130 **stretch of DNA on chromosome 4:** A. Puca et al., "A genome-wide scan for linkage to human exceptional longevity identifies a locus on chromosome 4," *PNAS* 98 (2001): 10505–8.

131 **Forming the company:** See Chesbrough, "Centagenetix: Building a Business Model for Genetic Longevity," Harvard Business School case study.

131 **Perls and colleagues reported:** B. J. Geesaman et al., "Haplotype-based identification of a microsomal transfer protein marker associated with the human lifespan," *PNAS* 100 (2003): 14115–20.

131 **Danish and German teams:** L. Bathum et al., "No evidence for an association between extreme longevity and Microsomal Transfer Protein polymorphisms in a longitudinal study of 1651 nonagenarians," *European Journal of Human Genetics* 13 (2005): 1154–58; and A. Nebel et al., "No association between microsomal triglyceride transfer protein (MTP) haplotype and longevity in humans," *PNAS* 102 (2005): 7906–9.

131 **By 2008 similar doubts:** See K. Christensen et al., "The quest for genetic determinants of human longevity: challenges and insights," *Nature Reviews Genetics* 7 (2006): 436–48.

132 **Barzilai has led the way:** See http://www.aecom.yu.edu.

133 **high levels of HDL:** Claudia Dreifus, "A Conversation with Nir Barzilai: It's Not the Yogurt; Looking for Longevity Genes," *New York Times*, February 24, 2004.

133 **protein called CETP:** See G. M. Martin et al., "Genetic Determinants of Human Health Span and Life Span: Progress and New Opportunities," *PLoS Genetics* 3 (2007): 1121–30.

133 **intriguing links:** Y. Suh et al., "Functionally significant insulin-like growth factor I receptor mutations in centenarians," *PNAS* 105 (2008): 3438–42.

133 **gene called FOXO3a:** B. J. Willcox et al., "FOXO3A gentotype is strongly associated with human longevity," *PNAS* 105 (2008): 13987–92.

134 **apologists warn:** See Kass, "L'Chaim and Its Limits: Why Not Immortality," *First Things*.

134 **Tithonus:** See Gruman, "A History of Ideas about the Prolongation of Life: The Evolution of Prolongevity Hypothesis to 1800," *Transactions of the American Philosophical Society*.

135 **cast doubt:** J. F. Fries, "Aging, Natural Death, and the Compression of Morbidity," *New England Journal of Medicine* 303 (1980): 130–35.

136 **chronically disabled dropped:** K. G. Manton et al., "Change in chronic disability from 1982 to 2004/2005 as measured by long-term changes in function and health in the U.S. elderly population," *PNAS* 103 (2006): 18374–79.

136 **review of disability:** G. Lafortune, G. Balestat, and the Disability Study Expert Group Members, 2007, "Trends in Severe Disability among Elderly People: Assessing the Evidence in 12 OECD Countries and Future Implications," *OECD Health Working Papers No. 26*, OECD.

137 **Fries has helped compile:** See J. F. Fries, "Frailty, Heart Disease, and Stroke: The Compression of Morbidity Paradigm," *American Journal of Preventive Medicine* 29 (2005): 164–68.

137 **Americans over eighty-five rose:** Manton et al., "Change in chronic disability from 1982 to 2004/2005 as measured by long-term changes in function and health in the U.S. elderly population," *PNAS*.

137 **Perls has reported:** See V. Mor and T. T. Perls, "Measuring Functional Decline in Population Aging in a Changing World and an Evolving Biology," *Journals of Gerontology Series A* 59 (2004): M609–11.

137 **prevalence of such illnesses:** See E. M. Crimmins, "Trends in the health of the elderly," *Annual Review of Public Health* 25 (2004): 79–98.

7: CRACKING THE LIFE-SPAN BARRIER

140 **his initial report:** C. M. McCay and M. F. Crowell, "Prolonging the Life Span," *Scientific Monthly* 39 (1934): 405–14.

140 **few researchers who did study:** See E. J. Masoro, "Subfield History: Calorie Restriction, Slowing Aging, and Extending Life," *Science of Aging Knowledge Environment,* February 26, 2003.

140 **drugs that mimic CR's effects:** See, for example, J. A. Baur et al., "Resveratrol improves health and survival of mice on a high-calorie diet," *Nature* 444 (2006): 337–42; R. Strong et al., "Nordihydroguaiaretic acid and aspirin increase lifespan of genetically heterogeneous male mice," *Aging Cell* 7 (2008): 641–50; and V. Wanke et al., "Caffeine extends yeast lifespan by targeting TORC1," *Molecular Microbiology* 69 (2008): 277–85.

140 **In 2005, the RAND:** D. P. Goldman et al., "Consequences of Health Trends and Medical Innovation for the Future Elderly," at http://content.healthaffairs.org.

141 **Born in 1898:** Biographical details on McCay drawn from J. K. Loosli, "Clive Maine McCay (1898–1967)—a biographical sketch," *Journal of Nutrition* 103 (1973): 1–10; and W. Shurtleff and A. Aoyagi, "Clive and Jeanette McCay, and The New York State Emergency Food Commission: Work with Soy," Soyinfo Center, unpublished manuscript.

142 **known for landmark studies:** T. B. Osborne et al., "The effect of retardation of growth upon the breeding period and duration of life in rats," *Science* 45 (1917): 294–95; and T. B. Osborne and L. B. Mendel, "The Resumption of Growth after Long Continued Failure to Grow," *Journal of Biological Chemistry* 23 (1915): 439–54.

142 **McCay asked his mentor:** Loosli, "Clive Maine McCay (1898–1967)—a biographical sketch," 4.

142 **fish on a low-protein:** C. M. McCay et al., "Growth Rates of Brook Trout Reared upon Purified Rations," *Journal of Nutrition* 1 (1929), 233.

143 **Cornaro's meager diet:** C. M. McCay, "Effect of Restricted Feeding upon

Aging and Chronic Diseases in Rats and Dogs," *American Journal of Public Health* 37 (1947): 521–28.

143 **immunologist Carlo Moreschi**: Ibid., 523.

143 **Peyton Rous**: P. Rous, "The influence of diet on transplant and spontaneous tumors," *Journal of Experimental Medicine* 20 (1914): 433–51.

143 **fast-growing mice live longer**: T. B. Robertson and L. A. Ray, "On the growth of relatively long lived compared with that of relatively short lived animals," *Journal of Biological Chemistry* 42 (1920), 71–107.

143 **Buffon theorized**: McCay, Crowell, "Prolonging the Life Span," 410.

143 **Researchers exploded it**: R. Weindruch and R. L. Walford, "Dietary restriction in mice beginning at 1 year of age: effects on lifespan and spontaneous cancer incidence," *Science* 215 (1982): 1415–18.

143 **McCay got a lot right**: C. M. McCay et al., "The Effect of Retarded Growth upon the Length of Life Span and upon the Ultimate Body Size," *Journal of Nutrition* 10 (1935): 63–79.

143 **food-restricted females died**: Ibid., 70.

144 **male data were as clear**: Ibid., 71.

144 **the following decade**: See C. M. McCay et al., "Experimental Prolongation of the Life Span," *Bulletin of the New York Academy of Medicine* 32 (1956): 91–101; and McCay, "Effect of Restricted Feeding upon Aging and Chronic Diseases in Rats and Dogs."

144 **1940s and after**: See McCay, "Effect of Restricted Feeding upon Aging and Chronic Diseases in Rats and Dogs"; and Loosli, "Clive Maine McCay (1898–1967)—a biographical sketch."

146 **life expectancy would climb**: J. R. Speakman and C. Hambly, "Starving for Life: What Animal Studies Can and Cannot Tell Us about the Use of Caloric Restriction to Prolong Human Lifespan," *Journal of Nutrition* 137 (2007): 1078–86.

146 **group of prominent gerontologists**: S. J. Olshansky et al., "The Longevity Dividend," *The Scientist* 20 (March 2006), 28.

146 **Mediterranean fruit flies**: J. R. Carey et al., "Life history response of Mediterranean fruit flies to dietary restriction," *Aging Cell* 1 (2002): 140–48.

146 **mice called DBA/2**: Finch, *The Biology of Human Longevity*, 227.

146 **list of CR responders**: E. J. Masoro, "Dietary restriction-induced life extension: a broadly based biological phenomenon," *Biogerontology* 7 (2006): 153–55.

146 **methionine**: See R. A. Miller et al., "Methionine-deficient diet extends mouse lifespan, slows immune and lens aging, alters glucose, T4, IGF-1 and insuline levels, and increases hepatocyte MIF levels and stress resistance," *Aging Cell* 4 (2005): 119–25.

147 **51 percent of elderly female**: R. T. Bronson and R. D. Lipman, "Reduction in Rate of Occurrence of Age Related Lesions in Dietary Restricted Laboratory Mice," *Growth, Development and Aging* 55 (1991): 169–84.

147 **suppress tumors**: Finch, *The Biology of Human Longevity*, 185.

147 **McCay reported**: McCay, "Effect of Restricted Feeding upon Aging and Chronic Diseases in Rats and Dogs," 525.

147 **representative rat study**: P. H. Duffy et al., "The effects of different levels of

dietary restriction on non-neoplastic diseases in male Sprague-Dawley rats," *Aging Clinical and Experimental Research* 16 (2004): 68–78.

147 **members of the Calorie Restriction Society**: T. E. Meyer et al., "Long-term caloric restriction ameliorates the decline in diastolic function in humans," *Journal of the American College of Cardiology* 47 (2006): 398–402.

147 **One of CR's most important**: Finch, *The Biology of Human Longevity*, 198.

147 **keeps back overzealous**: G. Fernandes et al., "Influence of diet on survival of mice," *PNAS* 73 (1976): 1279–83.

147 **immune-system weakening**: Finch, *The Biology of Human Longevity*, 192–97.

147 **rodent version of Alzheimer's disease**: Ibid., 219.

148 **look much younger**: McCay et al., "Experimental Prolongation of the Life Span," 96.

148 **hearing acuity**: S. Someya et al., "Caloric restriction suppresses apoptotic cell death in the mammalian cochlean and leads to prevention of presbycusis," *Neurobiology of Aging* 28 (2007): 1613–22.

148 **loss of muscle**: R. J. Colman et al., "Attenuation of sarcopenia by dietary restriction in rhesus monkeys," *Journals of Gerontology Series A.* 63 (2008): 556–59.

148 **a fourth of rodents on**: C. R. Finch, *The Biology of Human Longevity*, 185.

149 **drops in blood sugar**: Ibid., 186.

149 **described the mysterious deaths**: "Seek Key to Anti-aging in Calorie Cutback," *Wall Street Journal*, October 30, 2006, 1.

150 **St. Louis study**: See David Stipp, "Live a Lot Longer," *Fortune*, June 5, 1999, 144–60.

150 **much more interesting phenomenon**: A. R. Heydari et al., "Expression of Heat Shock Protein 70 Is Altered by Age and Diet at the Level of Transcription," *Molecular and Cellular Biology* 13 (1993): 2909–18.

151 **Johnson and colleagues postulated**: G. Lithgow et al., "Thermotolerance and extended life-span conferred by single-gene mutations and induced by thermal stress," *PNAS* 92 (1995): 7540–44.

151 **If you're a mouse**: Fernandes et al., "Influence of diet on survival of mice," *PNAS*.

151 **If you're a rat**: Duffy et al., "The effects of different levels of dietary restriction on non-neoplastic diseases in male Sprague-Dawley rats," *Aging Clinical and Experimental Research.*

151 **Masoro . . . strengthened**: E. J. Masoro, "Hormesis and the anti-aging effect of dietary restriction," *Experimental Gerontology* 33 (1998): 61–66.

152 **hormesis was an esteemed concept**: See E. J. Calabrese and L. A. Baldwin, "Chemical hormesis: its historical foundations as a biological hypothesis," *Human & Experimental Toxicology* 19 (2000): 2–31; and David Stipp, "A Little Poison Can Be Good for You," *Fortune*, May 28, 2003.

153 **benefits of consuming lots of vegetables**: See D. Gems and L. Partridge, "Stress-Response Hormesis and Aging: "That Which Does Not Kill Us Makes Us Stronger," *Cell Metabolism* 7 (2008): 200–203.

153 **have made a compelling case**: See S. Rattan, "Applying hormesis in aging research and therapy," *Human & Experimental Toxicology* 20 (2001): 281–86.

153 **Hormesis isn't all**: E. J. Masoro, "Caloric restriction-induced life extension of rats and mice: A critique of proposed mechanisms," *Biochimica et Biophysica Acta* (2009): 1–9.

153 **"starvation response"**: D. E. Harrison and J. R. Archer, "Natural selection for extended longevity from food restriction," *Growth Development and Aging* 52 (1988): 65; and R. Holliday, "Food, reproduction and longevity—is the extended lifespan of calorie-restricted animals an evolutionary adaptation?" *Bioessays* 10 (1989): 125–27.

154 **alternative theory**: Personal communication, Steven Austad.

154 **run themselves to death**: See Finch, *The Biology of Human Longevity*, 230; and A. A. van Elburg et al., "Nurse evaluation of hyperactivity in anorexia nervosa: a comparative study," *European Eating Disorders Review* 15 (2007): 425–29.

155 **normal mice on CR and long-lived dwarf mice**: V. D. Longo and C. E. Finch, "Evolutionary medicine: from dwarf model systems to healthy centenarians?" *Science* 299 (2003): 1343–46.

155 **low insulin and body temperature**: G. S. Roth et al., "Biomarkers of caloric restriction may predict longevity in humans," *Science* 297 (2002): 811.

155 **Different methods of restricting**: E. L. Greer and A. Brunet, "Different dietary restriction regimens extend lifespan by both independent and overlapping genetic pathways in C. elegans," *Aging Cell* 8 (2009): 113–27.

155 **already-long lives**: M. M. Masternak et al., "Divergent effects of calorie restriction on gene expression in normal and long-lived mice," *Journals of Gerontology Series A* 59 (2004): 784–88.

156 **CR's well-known downsides**: See http://www.calorierestriction.org; E. M. Gardner, "Caloric restriction decreases survival of aged mice in response to primary influenza infection," *Journals of Gerontology Series A* (2005): 688–94; and B. W. Ritz et al., "Energy restriction impairs natural killer cell function and increases the severity of influenza infection in young adult male C57BL/6 mice," *Journal of Nutrition* 138 (2008): 2269–75.

156 **driven to cannibalism**: McCay, "Effect of Restricted Feeding upon Aging and Chronic Diseases in Rats and Dogs," 524.

156 **including relentless hunger**: Speakman and Hambly, "Starving for Life: What Animal Studies Can and Cannot Tell Us about the Use of Caloric Restriction to Prolong Human Lifespan," 1081.

157 **Leonard Hayflick, for example**: Leonard Hayflick, *How and Why We Age* (New York: Ballantine Books, 1994), 277–95.

157 **Austad has found**: See S. N. Austad and D. M. Kristan, "Are mice calorically restricted in nature?" *Aging Cell* 2 (2003): 201–7; and J. M. Harper et al., "Does caloric restriction extend life in wild mice?" *Aging Cell* 5 (2006): 441–49.

157 **pessimists' best shot**: See E. Le Bourg, "Dietary Restriction would probably not increase longevity in human beings and other species able to leave unsuitable environments," *Biogerontology* 7 (2006): 149–52; R. Holliday, "Food, fertility and longevity," *Biogerontology* 7 (2006): 139–41; D. P. Shanley and T. B. L. Kirkwood, "Caloric restriction does not enhance longevity in all species and is unlikely to do so in humans," *Biogerontology* 7 (2006): 165–

68; and J. P. Phelan and M. R. Rose, "Caloric restriction increases longevity substantially only when the reaction norm is steep," *Biogerontology* 7 (2006): 161–64.

157 **Optimists counter**: See Masoro, "Dietary restriction-induced life extension: a broadly based biological phenomenon."

157 **case for optimism**: See R. Weindruch, "Will dietary restriction work in primates?" *Biogerontology* 7 (2006): 169–71; and D. K. Ingram et al., "The potential for dietary restriction to increase longevity in humans: extrapolation from monkey studies," *Biogerontology* 7 (2006): 143–48.

158 **Weindruch's group made a splash**: R. J. Colman et al., "Caloric restriction delays disease onset and mortality in rhesus monkeys," *Science* 325 (2009): 201–4.

158 **Correlates of slowed aging**: See Meyer et al., "Long-term caloric restriction ameliorates the decline in diastolic function in humans"; and L. K. Heilbronn et al., "Effect of 6-month calorie restriction on biomarkers of longevity, metabolic adaptation, and oxidative stress in overweight individuals: a randomized controlled trial," *Journal of the American Medical Association* 295 (2006): 1539–48.

159 **Roy Walford**: Biographical details on Walford drawn from Thomas H. Maugh II, "Obituary: Roy Walford," 79; Eccentric UCLA Scientist Touted Food Restriction," *Los Angeles Times*, May 1, 2004; and Gary Taubes, "Staying Alive," *Discover*, February 2000.

159 **Richard Miller points out**: Personal communication, Richard Miller.

159 **biosphereans' hormonal**: R. L. Walford et al., "Calorie Restriction in Biosphere 2: Alterations in Physiologic, Hematologic, Hormonal, and Biochemical Parameters in Humans Restricted for a 2-Year Period," *Journals of Gerontology Series A* 57 (2002): B211–24.

160 **longest-living 1 percent of Okinawans**: D. C. Willcox et al., "Caloric restriction and human longevity: What can we learn from the Okinawans?" *Biogerontology* 7 (2006): 173–77.

161 **Okinawans also appear**: See Okinawa Centenarian Study Web site at http://www.okicent.org.

8: THE GREAT FREE LUNCH

162 **first significant attempt**: Personal communications, Donald Ingram and George Roth; and George S. Roth, *The Truth About Aging: Can We Really Live Longer and Healthier?* (Port Orchard, WA: Windstorm Creative, 2005).

163 **2-deoxy-D-glucose**: See M. Rezek and E. A. Kroeger, "Glucose Antimetabolites and Hunger," *Journal of Nutrition* 106 (1976): 143–57.

163 **2DG's physiological effects**: See D. K. Ingram et al., "Calorie restriction mimetics: an emerging research field," *Aging Cell* 5 (2006): 97–108; and M. A. Lane et al., "2-Deoxy-D-Glucose Feeding in Rats Mimics Physiologic Effects of Calorie Restriction," *Journal of Anti-Aging Medicine* 1 (1998): 327–37.

163 **who'd taken 2DG**: Roth, *The Truth About Aging*, 171.

164 **initial finding**: Ibid., 170.

164 **study's overall results**: Lane et al., "2-Deoxy-D-Glucose Feeding in Rats Mimics Physiologic Effects of Calorie Restriction," *Journal of Anti-Aging Medicine.*

165 **seemed that the medical world's**: D. K. Ingram et al., "Calorie restriction mimetics: an emerging research field," *Aging Cell*; and W. Duan and M. P. Mattson, "Dietary restriction and 2-deoxyglucose administration improve behavioral outcome and reduce degeneration of dopaminergic neurons in models of Parkinson's disease," *Journal of Neuroscience Research* 57 (1999): 195–206.

165 **"the millennium experiment"**: Roth, *The Truth About Aging*, 177.

165 **"began to kill rats"**: Ibid., 177.

166 **in 2009 Roth disclosed**: See Janet Raloff, "Coming: Ersatz calorie restriction; Avocados may hold a key to longer, better health," *Science News*, April 20, 2009.

168 **share two telltale traits**: G. S. Roth et al., "Biomarkers of caloric restriction may predict longevity in humans," *Science* 297 (2002): 811.

169 **raise some knotty issues**: See R. A. Miller et al., "Interpretation, design, and analysis of gene array expression experiments," *Journals of Gerontology Series A* 56 (2001): B52–57.

169 **chips have very intriguing things**: See C. K. Lee et al., "Transcriptional profiles associated with aging and middle age-onset caloric restriction in mouse hearts," *PNAS* 99 (2002): 14988–93; C. K. Lee et al., "Gene-expression profile of the ageing brain in mice," *Nature Genetics* 25 (2000): 294–97; J. M. Dhahbi et al., "Temporal linkage between the phenotypic and genomic responses to caloric restriction," *PNAS* 101 (2004): 5524–29; S. X. Cao et al., "Genomic profiling of short- and long-term caloric restriction effects in the liver of aging mice," *PNAS* 98 (2001): 10630–35; and S. R. Spindler, "Use of microarray biomarkers to identify longevity therapeutics," *Aging Cell* 5 (2006): 39–50.

169 **changes in mere weeks**: See S. R. Spindler, "Use of microarray biomarkers to identify longevity therapeutics," *Aging Cell.*

169 **repair preexisting damage**: Ibid.

169 **metformin is remarkably good**: J. M. Dhahbi et al., "Identification of potential caloric restriction mimetics by microarray profiling," *Physiological Genomics* 23 (2005): 343–50.

169 **French lilac**: L. A .Witters, "The Blooming of the French Lilac," *Journal of Clinical Investigation* 108 (2001): 1105–7.

170 **increase their average life span by 23 percent**: V. M. Dilman and V. N. Anisimov, "Effect of treatment with phenformin, diphenylhydantoin or L-dopa on life span and tumour incidence in C2H/Sn mice," *Gerontology* 26 (1980): 241–46.

170 **The Russians' 2008 report**: V. N. Anisimov et al., "Metformin slows down aging and extends life span of female SHR mice," *Cell Cycle* 7 (2008): 2769–73.

171 **enthusiasts are already blogging**: See http://www.lef.org.

172 **Guarente's career**: Biographical details on Guarente drawn from Lenny

Guarente, *Ageless Quest: One Scientist's Search for Genes That Prolong Youth* (Cold Spring Harbor, NY: Cold Spring Harbor Press, 2003); and from personal communication, Lenny Guarente.

173 **bud off daughter cells**: See R. Mortimer and J. Johnson, "Life span of individual yeast cells," *Nature* 183 (1959): 1751–52; and K. J. Bitterman et al., "Longevity Regulation in *Saccharomyces cerevisiae*: Linking Metabolism, Genome Stability, and Heterochromatin," *Microbiology and Molecular Biology Reviews* 67 (2003): 376–99.

176 **doubly serendipitous**: See B. K. Kennedy et al., "Mutation in the Silencing Gene SIR4 Can Delay Aging in S. cerevisiae," *Cell* 80 (1995): 485–96.

176 **variant of a previously identified**: Ibid.

177 **Sinclair met Guarente**: Personal communication, David Sinclair.

178 **one of Sinclair's first acts**: Guarente, *Ageless Quest*, 35.

178 **Sinclair flipped open**: Personal communication, David Sinclair.

178 **form little circles**: D. A. Sinclair and L. Guarente, "Extrachromosomal rDNA Circles—A Cause of Aging in Yeast," *Cell* 91 (1997): 1033–42.

179 **singled out SIR2**: M. Kaeberlein et al., "The SIR2/3/4 complex and SIR2 alone promote longevity in Saccharomyces cerevisiae by two different mechanisms," *Genes & Development* 13 (1999): 2570–80.

179 **rivals had stolen**: Guarente, *Ageless Quest*, 49–52.

180 **Imai had shown**: S. Imai et al., "Transcriptional silencing and longevity protein Sir2 is an NAD-dependent histone deacetylase," *Nature* 403 (2000): 295–300.

181 **Michael West**: See David Stipp, "The Hunt for the Youth Pill: From Cell-Immortalizing Drugs to Cloned Organs, Biotech Finds New Ways to Fight Against Time's Toll," *Fortune*, October 11, 1999.

181 **Guarente found evidence in yeast**: Guarente, *Ageless Quest*, 41.

182 **Geron deemphasized**: See Stipp, "The Hunt for the Youth Pill: From Cell-Immortalizing Drugs to Cloned Organs, Biotech Finds New Ways to Fight Against Time's Toll," *Fortune*.

182 **mice age and die**: See S. Kim et al., "Telomeres, aging and cancer: In search of a happy ending," *Oncogene* 21 (2002): 503–11.

182 **Spanish scientists reported**: A. Tomas-Loba et al., "Telomerase Reverse Transcriptase Delays Aging in Cancer-Resistant Mice," *Cell* 135 (2008): 609–22.

182 **centenarians carry variants of the telomerase gene:** G. Atzmon et al., "Genetic variation in human telomerase is associated with telomere length in Ashkenazi centenarians," *PNAS* (epub, December 4, 2009).

182 **Su-Ju Lin**: S. J. Lin et al., "Requirement of NAD and SIR2 for life-span extension by caloric restriction in Saccharomyces cerevisiae," *Science* 289 (2000): 2126–28.

183 **Heidi Tissenbaum**: H. A. Tissenbaum and L. Guarente, "Increased dosage of a sir-2 gene extends lifespan in Caenorhabditis elegans," *Nature* 410 (2001): 227–30.

183 **Guarente and Kenyon coauthored**: L. Guarente and C. Kenyon, "Genetic pathways that regulate ageing in model organisms," *Nature* 408 (2000): 255–62.

183 **resonant chord with Jonathan Fleming**: Guarente, *Ageless Quest,* 115–24.

184 **start-up gained momentum**: See Solomon, *The Quest for Human Longevity,* 95–112.

184 **Elixir's merger with Centagenetix**: Ibid.

184 **ran a lengthy feature**: James Burnett, "Take This Pill and Live Forever," *Boston,* April 2003.

9: RED WINE'S ENIGMATIC DIRTY DRUG

185 **from whose life**: Biographical details drawn from personal communication, David Sinclair.

186 **questioned his former mentor's**: See J. Couzin, "Aging research's family feud," *Science* 303 (2004): 1276–79.

186 **splash by identifying resveratrol**: K. T. Howitz et al., "Small molecule activators of sirtuins extend Saccharomyces cerevisiae lifespan," *Nature* 425 (2003): 191–96.

187 **told *Science* magazine**: See Couzin, "Aging research's family feud," *Science.*

187 **Harvard Medical School**: See http://hms.harvard.edu.

187 **Its first step**: Personal communication, David Sinclair.

188 **piceatannol and quercetin**: See Howitz et al., "Small molecule activators of sirtuins extend Saccharomyces cerevisiae lifespan," *Nature.*

189 **First isolated in 1940**: See J. A. Baur and D. A. Sinclair, "Therapeutic potential of resveratrol: the in vivo evidence," *Nature Reviews* 5 (2006): 493–506.

189 **food scientists suggested**: E. H. Siemann and L. L. Creasy, "Concentration of the phytoalexin resveratrol in wine," *American Journal of Enology and Viticulture* 43 (1992): 49–52.

189 **popular news show**: See Hilary Abramson, "The Flip Side of French Drinking," *The Marin Institute* (Winter 2000).

189 **proposed a sounder theory**: M. Jan et al., "Cancer chemopreventive activity of resveratrol, a natural product derived from grapes," *Science* 275 (1997): 218–20.

189 **rose at an exponential rate**: See Baur and Sinclair, "Therapeutic potential of resveratrol: the in vivo evidence," *Nature Reviews.*

189 **carrying it out himself**: Personal communication, David Sinclair.

190 ***Nature* published the study**: Howitz et al., "Small molecule activators of sirtuins extend Saccharomyces cerevisiae lifespan," *Nature.*

191 **Sinclair proved himself**: See Mary Carmichael, "Chemical Wowie," *Newsweek,* September 1, 2003, 9; Stephen Smith, "In Lab, Seeking Secret of Youth Chemical Abundant in Red Wine Appears to Slow Aging in Study," *Boston Globe,* August 25, 2003, A2; and "Better red than dead, study finds," *Daily Telegraph,* August 26, 2003, 5.

191 **Bill Sardi**: See Bill Sardi, *The Red Wine Pill: The Miracle of Resveratrol* (San Dimas, CA: Here and Now Books, 2004).

191 **Sardi briefly enlisted**: See Couzin, "Aging research's family feud," *Science.*

192 **resveratrol promiscuously consorts**: See Baur and Sinclair, "Therapeutic potential of resveratrol: the in vivo evidence," *Nature Reviews.*

193 **cast doubt on some of the Guarente lab's:** Personal communication, Matt Kaeberlein.

193 **a 2005 paper coauthored:** M. Kaeberlein et al., "Substrate-specific activation of sirtuins by resveratrol," *Journal of Biology and Chemistry* 280 (2005): 17038–45.

194 **a similar study:** M. T. Borra et al., "Mechanism of human SIRT1 activation by resveratrol," *Journal of Biology and Chemistry* 280 (2005): 17187–95.

194 **critiques went way beyond:** See M. Kaeberlein et al., "Sir2-independent life span extension by calorie restriction in yeast," *PLoS Biology* 2 (2004): E296; and M. Kaeberlein and B. K. Kennedy, "Large-scale identification in yeast of conserved ageing genes," *Mechanisms of Ageing and Development* 126 (2005): 17–21.

195 **fruit flies and nematodes:** J. G. Wood et al., "Sirtuin activators mimic caloric restriction and delay ageing in metazoans," *Nature* 430 (2004): 686–89.

195 **is absent in mice:** D. Chen et al., "Increase in activity during calorie restriction requires Sirt1," *Science* 310 (2005): 1641.

195 **The MIT group also reported:** F. Picard et al., "Sirt1 promotes fat mobilization in white adipocytes by repressing PPAR-gamma," *Nature* 429 (2004): 771–76.

195 **A collaborative study:** J. Luo et al., "Negative control of p53 by Sir2alpha promotes cell survival under stress," *Cell* 107 (2001): 137–48.

195 **helps prevent neurons:** T. Araki et al., "Increased nuclear NAD biosynthesis and SIRT1 activation prevent axonal degeneration," *Science* 305 (2004): 1010–13.

195 **SIRT1 in the heart cells:** R. R. Alcendor et al., "Sirt1 regulations aging and resistance to oxidative stress in the heart," *Circulation Research* 100 (2007): 1512–21.

195 **Molecular details:** A. Brunet et al., "Stress-dependent regulation of FOXO transcription factors by the SIRT1 deacetylase," *Science* 303 (2004): 2011–15.

196 **SIRT1 inhibits p53:** See L. Guarente, "Calorie restriction and SIR2 genes— towards a mechanism," *Mechanisms of Ageing and Development.* 126 (2005): 923–28.

196 **induce cells' resistance to stress:** S. D. Westerheide et al., "Stress-inducible regulation of heat shock factor 1 by the Deacetylase SIRT1," *Science* 323 (2009): 1063–66.

196 **Kaeberlein didn't take:** See I. Chen, "Rookie rising," *Science of Aging Knowledge Environment*, December 18, 2002.

197 **By 2006, the two sides:** See J. Rine, "Twists in the tale of the aging yeast," *Science* 310 (2005): 1124–25.

197 **turquoise killifish:** D. R. Valenzano et al., "Resveratrol prolongs lifespan and retards the onset of age-related markers in a short-lived vertebrate," *Current Biology* 16 (2006): 296–300.

198 **man named Tom LoGiudice:** Personal communication, David Sinclair.

200 **the study's initial findings:** J. A. Baur et al., "Resveratrol improves health and survival of mice on a high-calorie diet," *Nature* 444 (2006): 337–42.

200 ***New York Times* headlined:** Nicholas Wade, "Red Wine Holds Answer. Check Dosage," *New York Times,* November 2, 2006.

201 **Sirtris was already making fast progress:** See J. C. Milne et al., "Small molecule activators of SIRT1 as therapeutics for the treatment of type 2 diabetes," *Nature* 450 (2007): 712–16.

202 **French team's data:** M. Lagouge et al., "Resveratrol improves mitochondrial function and protects against metabolic disease by activating SIRT1 and PGC-1alpha," *Cell* 127 (2006): 1109–22.

202 **PGC-1-alpha:** See J. T. Rodgers et al., "Nutrient control of glucose homeostasis through a complex of PGC-1alpha and SIRT1," *Nature* 434 (2005): 113-18.

10: RAPAMYCIN AND THE TALE OF TOR

204 **resveratrol did not extend:** K. J. Pearson et al., "Resveratrol delays age-related deterioration and mimics transcriptional aspects of dietary restriction without extending life span," *Cell Metabolism* 8 (2008): 157–68.

205 **jointly reported in** *Nature:* D. E Harrison et al., "Rapamycin fed late in life extends lifespan in genetically heterogeneous mice," *Nature* 460 (2009): 392–95.

205 **lives were actually shortened:** M. J. Forster et al., "Genotype and age influence the effect of caloric intake on mortality in mice," *Federation of American Societies for Experimental Biology Journal* 17 (2003): 690–92.

205 **veterans had little hope:** See, for example, W. Sansom, "HSC scientists discover mice live longer with rapamycin," University of Texas Health Science Center news release, July 13, 2009.

205 **It began in 1964:** See Barry D. Kahan, "Discoverer of the treasure from a barren island: Suren Sehgal (10 February 1932 to 21 January 2003)," *Transplantion* 76 (2003): 623–24.

206 **Selman Waksman:** See http://waksman.rutgers.edu.

206 **there is now evidence:** C. Mlot, "Antibiotics in nature: beyond biological warfare," *Science* 324 (2009): 1637–39.

207 **isolated an antifungal compound:** See Kahan, "Discoverer of the treasure from a barren island: Suren Sehgal (10 February 1932 to 21 January 2003)," *Transplantion.*

207 **antitumor effects:** S. Huang et al., "Rapamycins: Mechanism of Action and Cellular Resistance," *Cancer Biology & Therapy* 2 (2003): 222–32.

207 **Hall's group discovered:** J. Heitman et al., "Targets for cell cycle arrest by the immunosuppressant rapamycin in yeast," *Science* 23 (1991): 905–9.

207 **worms, flies, mammals, and even plants:** See J. L. Crespo and M. N. Hall, "Elucidating TOR Signaling and Rapamycin Action: Lessons from Saccharomyces cerevisiae," *Microbiology & Molecular Biology Reviews* 66 (2002): 579–91.

207 **By 2000 it had become clear:** Ibid.

207 **Japanese scientists discovered:** See T. Noda and Y. Ohsumi, "Tor, a phosphatidylinositol kinase homologue, controls autophagy in yeast," *Journal of Biology and Chemistry* 273 (1998): 3963–66.

207 **boost cells' defenses:** See Crespo and Hall, "Elucidating TOR Signaling and Rapamycin Action: Lessons from Saccharomyces cerevisiae," *Microbiology & Molecular Biology Reviews.*

207 **especially intriguing discovery**: Ibid.

208 **Sharp got interested**: Personal communication, Dave Sharp.

208 **study with Andrzej Bartke**: Z. D. Sharp and A. Bartke, "Evidence for down-regulation of phosphoinositide 3-kinase/Akt/mammalian target of rapamycin (PI3K/Akt/mTOR)-dependent translation regulatory signaling pathways in Ames dwarf mice," *Journals of Gerontology Series A* 60 (2005): 293–300.

209 **suppressing TOR in nematodes**: T. Vellai et al., "Influence of TOR kinase on lifespan in C. elegans," *Nature* 426 (2003): 620.

209 **turning down TOR in fruit flies**: P. Kapahi et al., "Regulation of lifespan in Drosophila by modulation of genes in the TOR signaling pathway," *Current Biology* 14 (2004): 885–90.

209 **loud wake-up call**: M. Kaeberlein et al., "Regulation of yeast replicative life span by TOR and Sch9 in response to nutrients," *Science* 310 (2005): 1193–96.

209 **Topping TOR's to-do list**: See Crespo and Hall, "Elucidating TOR Signaling and Rapamycin Action: Lessons from Saccharomyces cerevisiae," *Microbiology & Molecular Biology Reviews*; and M. Hansen et al., "Lifespan extension by conditions that inhibit translation in Caenorhabditis elegans," *Aging Cell* 6 (2007): 95–110.

210 **A number of chemical weapons**: Personal communication, Gary Ruvkun.

210 **another benefit of scaling back**: See T. Vellai, "Autophagy genes and ageing," *Cell Death and Differentiation* 16 (2009): 94–102.

210 **proteins can form toxic aggregates**: See W. E. Balch et al., "Adapting proteostasis for disease intervention," *Science* 319 (2008): 916–19.

210 **Stimulating autophagy**: See Vellai, "Autophagy genes and ageing," *Cell Death and Differentiation*.

211 **mammals rely on autophagy**: A. Kuma et al., "The role of autophagy during the early neonatal starvation period," *Nature* 432 (2004): 1032–36.

211 **he formally proposed**: Personal communication, Dave Sharp.

211 **ITP itself dates**: Personal communication, Donald Ingram.

212 **the ITP launches**: See R. A. Miller et al., "An aging interventions testing program: study design and interim report," *Aging Cell* 6 (2007): 565–75.

212 **first round**: R. Strong et al., "Nordihydroguaiaretic acid and aspirin increase lifespan of genetically heterogeneous male mice," *Aging Cell* 7 (2008): 641–50.

212 **both inhibit TOR**: Personal communication, Randy Strong.

213 **Before testing rapamycin**: Ibid.

213 **stuff worked like magic**: Harrison et al., "Rapamycin fed late in life extends lifespan in genetically heterogeneous mice," *Nature*.

214 **for starters, rapamycin can lower**: See M. V. Blagosklonny, "An anti-aging drug today: from senescence-promoting genes to anti-aging pill," *Drug Discovery Today* 12 (2007): 218–24; M. N. Stanfel et al., "The TOR pathway comes of age," *Biochimica et Biophysica Acta*, (June 15, 2009); and M. Kaeberlein and B. K. Kennedy, "Protein translation, 2008," *Aging Cell* 7 (2008): 777–82.

215 **hinder formation of new neuronal**: See B. Raught et al., "The target of rapamycin (TOR) proteins," *PNAS* 98 (2001): 7037–44.

215 **cholesterol and triglycerides**: See B. D. Kahan, "Fifteen years of clinical

studies and clinical practice in renal transplantation: reviewing outcomes with de novo use of sirolimus in combination with cyclosporine," *Transplantation Proceedings* 40 (2008): S17–20.

215 **clinical studies with rapamycin:** Ibid.

215 **rapamycin can ward off cancer:** See D. A. Guertin and D. M. Sabatini, "Defining the role of mTOR in cancer," *Cancer Cell* 12 (2007): 9–22.

215 **proteins that the AIDS virus:** See A. Heredia et al., "Rapamycin causes down-regulation of CCR5 and accumulation of anti-HIV Ð-chemokines: An approach to suppress R5 strains of HIV-1," *PNAS* 100 (2003): 10411–16.

216 **"memory T-cells":** K. Araki et al., "mTOR regulates memory CD8 T-cell differentiation," *Nature* (June 21, 2009).

216 **a 2004 cancer study:** E. Raymond et al., "Safety and Pharmacokinetics of Escalated Doses of Weekly Intravenous Infusion of CCI-779, A Novel mTOR Inhibitor, in Patients with Cancer," *Journal of Clinical Oncology* 22 (2004): 2336–47.

216 **seventy-six heart patients:** A. E. Rodriguez et al., "Role of oral rapamycin to prevent restenosis in patients with de novo lesions undergoing coronary stenting: results of the Argentina single centre study (ORAR trial)," *Heart* 91 (2005): 1433–37.

216 **CR itself might be characterized:** See Finch, *The Biology of Human Longevity,* 190–92.

217 **TOR-regulated enzyme, called S6K1:** C. Selman et al., "Ribosomal protein S6 kinase 1 signaling regulates mammalian life span," *Science* 326 (2009): 140–44.

217 **stimulation of AMPK:** C. Canto et al., "AMPK regulates energy expenditure by modulating NAD+ metabolism and SIRT1 activity," *Nature* 458 (2009): 1056–60.

217 **resveratrol's gene-activity fingerprint:** J. L. Barger et al., "A low dose of dietary resveratrol partially mimics caloric restriction and retards aging parameters in mice," *PLoS One* 3 (2008): 2264.

217 **mice implanted with extra SIRT1:** See L. Bordone et al., "SIRT1 transgenic mice show phenotypes resembling calorie restriction," *Aging Cell* 6 (2007): 759–67; and A. S. Banks et al., "SirT1 gain of function increases energy efficiency and prevents diabetes in mice," *Cell Metabolism* 8 (2008): 333–41.

217 **2007 study by Linda Partridge:** T. M. Bass et al., "Effects of resveratrol on lifespan in Drosophila melanogaster and Caenorhabditis elegans," *Mechanisms of Ageing and Development* 128 (2007): 546–52.

218 **test-tube studies in late 2009 and early 2010:** D. Beher et al., "Resveratrol is not a direct activator of SIRT1 enzyme activity," *Chemical Biology and Drug Design* 74 (2009): 619–24; M. Pacholec et al., "SRT1720, SRT2183, SRT1460, and resveratrol are not direct activators of SIRT1," *Journal of Biological Chemistry* (epub: January 8, 2010).

218 **reduce the chronic, low-level inflammation:** See S. Lavu et al., "Sirtuins—novel therapeutic targets to treat age-associated diseases," *Nature Reviews Drug Discovery* 7 (2008): 841–53.

218 **compelling cardiac benefits:** See J. L. Barger et al., "Short-term consumption

of a resveratrol-containing nutraceutical mixture mimics gene expression of long-term caloric restriction in mouse heart," *Experimental Gerontology* 43 (2008): 859–66.

218 **overlapping modes of action:** See O. Medvedik et al., "MSN2 and MSN4 Link Calorie Restriction and TOR to Sirtuin-Mediated Lifespan Extension in Sacccharomyces cerevisiae," *PLoS Biology* 5 (2007): 2330–41; I. H. Lee et al., "A role for the NAD-dependent deacetylase Sirt1 in the regulation of autophagy," *PNAS* 105 (2008): 3374–79; and Blagosklonny, "An anti-aging drug today: from senescence-promoting genes to anti-aging pill," *Drug Discovery Today*.

218 **compound is remarkably benign:** See, for example, L. D. Williams et al., "Safety studies conducted on high-purity trans-resveratrol in experimental animals," *Food and Chemical Toxicology* (epub, June 6, 2009).

219 **One notable risk:** See, for example, E. J. Kim and S. J. Um, "SIRT1: roles in aging and cancer," *Biochemistry and Molecular Biology Reports* 41 (2008): 751–56.

219 **resveratrol fights cancer:** See G. Boily et al., "SirT1-null mice develop tumors at normal rates but are poorly protected by resveratrol," *Oncogene* (2009), epub ahead of print; R. Firestein et al., "The SIRT1 deacetylase suppresses intestinal tumorigenesis and colon cancer growth," *PLoS One* 3 (2008): e2020; R. H. Wang et al., "Interplay among BRCA1, SIRT1, and Survivin during BRCA1-associated tumorigenesis," *Molecular Cell* 32 (2008): 11–20; and Lavu et al., "Sirtuins—novel therapeutic targets to treat age-associated diseases," *Nature Reviews Drug Discovery*.

219 **SIRT1's anti-aging magic:** P. Oberdoerffer et al., "SIRT1 redistribution on chromatin promotes genomic stability but alters gene expression during aging," *Cell* 135 (2008): 907–18.

220 **resveratrol is found in many foods:** See Joseph Maroon, *The Longevity Factor: How Resveratrol and Red Wine Activate Genes for a Longer and Healthier Life* (New York: Atria Books, 2009).

220 **For a rapamycinlike jolt:** See V. Wanke et al., "Caffeine extends yeast lifespan by targeting TORC1," *Molecular Microbiology* 69 (2008): 277–85.

220 **studies suggesting that aspirin:** See, for example, Z. Guo et al., "Aspirin Inhibits Serine Phosphorylation of Insulin Receptor Substate 1 in Tumor Necrosis Factor-treated Cells through Targeting Multiple Serine Kinases," *Journal of Biology and Chemistry* 278 (2003): 24944–50.

220 **Richard Miller has observed:** R. A. Miller, "Extending life: scientific prospects and political obstacles," *Milbank Quarterly* 80 (2002): 155–74.

11: SIRTRIS, MASTER VOYAGER OF THE VORTEX

222 **Westphal had begun the year:** Unless otherwise noted, the information in this chapter was drawn from reporting I did as a frequent visitor at Sirtris between 2005 and 2008.

223 **article in the *Wall Street Journal*:** David Hamilton, "Biotech Start-Ups Increasingly Opt for a Sale to Drug Firms over an IPO," *Wall Street Journal*, July 13, 2006, C1.

224 **a quiet, bookish youngster**: Personal communications, Heiner and Frauke Westphal.

231 **a whopping $113.5 million**: Sirtris Pharmaceuticals initial public offering prospectus, May 8, 2007, 45.

236 **licensed exclusive rights**: "Sirtris Licenses Fundamental Patent from Guarente Lab at MIT: SIRT1 Activation Has Therapeutic Potential for Lipid Disorders," Sirtris Pharmaceuticals press release, October 11, 2007.

236 **scores of diabetes patients**: See "Sirtris Announces Clinical Results from Phase 1 Trials of SRT501, a Sirtuin Therapeutic, Company Initiates Phase 1b Study in Type 2 Diabetes Patients," Sirtris Pharmaceuticals press release, October 4, 2006.

237 *Nature* **paper appeared**: J. C. Milne et al., "Small molecule activators of SIRT1 as therapeutics for the treatment of type 2 diabetes," *Nature* 450 (2007): 712–16.

238 **Auwerx chimed in**: J. N.Feige et al., "Specific SIRT1 Activation Mimics Low Energy Levels and Protects against Diet-Induced Metabolic Disorders by Enhancing Fat Oxidation," *Cell Metabolism* 8 (2008): 347–58.

238 **a group at Pfizer reported**: M. Pacholec, "SIRT1720, SRT2183, SRT1460, and resveratrol are not direct activators of SIRT1," *Journal of Biological Chemistry*.

238 **sirtuins have very complex functions**: See S. Lavu et al., "Sirtuins—novel therapeutic targets to treat age-associated diseases," *Nature Reviews* 7 (2008): 841–53; H. Yamamoto et al., "Sirtuin Functions in Health and Disease," *Molecular Endocrinology* 21 (2007): 1745–55; M. C. Haigis and L. P. Guarente, "Mammalian sirtuins—emerging roles in physiology, aging, and calorie restriction," *Genes & Development* 20 (2006): 2913–21; and W. C. Hallows et al., "Ure(k)a! Sirtuins Regulate Mitochondria," *Cell* 137 (2009): 404–6.

240 **study linking Avandia**: S. E. Nissen and K. Wolski, "Effect of Rosiglitazone on the Risk of Myocardial Infarction and Death from Cardiovascular Causes," *New England Journal of Medicine* 356 (2007): 2457–71.

240 **data on the issue**: See "The Rosiglitazone Story—Lessons from an FDA Advisory Committee Meeting," *New England Journal of Medicine* 357 (2007): 844.

241 **Avandia's sales plunged**: See Elena Berton, "Falling Avandia Sales Hit GlaxoSmithKline," *Dow Jones Newswire*, February 7, 2008.

241 **cost of developing a drug**: A. J. J. Wood, "A Proposal for Radical Changes in the Drug-Approval Process," *New England Journal of Medicine* 355 (2006): 618–23.

241 **As Witty commented**: See T. Stuart and J. Weber, "GSK's Acquisition of Sirtris: Independence or Integration," *Harvard Business School Case Study* N9-809-026, April 13, 2009.

243 **Elixir withdrew a planned IPO**: See Roxanne Palmer, "Elixir Raises $12 million, Says Drug Trial Could Lead to Novartis Purchase," xconomy.com, May 19, 2009.

12: THE GEORGE BURNS SCENARIO

245 **self-reported happiness**: See S. T. Charles and L. L. Carstensen, "Emotion Regulation and Aging," in *Handbook of Emotion Regulation*, ed. J. J. Gross (New York: Guilford Press, 2007).

246 **Thomas Lynch**: Thomas Lynch, "Why Buy More Time?" *New York Times,* March 14, 1999.

246 **a kind of apologist manifesto**: L. R. Kass, "L'Chaim and Its Limits: Why Not Immortality?" *First Things* 113 (2001): 17–24.

246 **council mirrored Kass's views**: The President's Council on Bioethics, "Beyond Therapy: Biotechnology and the Pursuit of Happiness," October 2003.

247 **"engineered negligible senescence"**: See http://www.sens.org.

248 **Kass wrote**: Kass, "L'Chaim and Its Limits: Why Not Immortality?" *First Things.*

249 **British bioethicist John Harris**: J. Harris, "Immortal ethics," *Annals of the New York Academy of Sciences* 1019 (2004): 527–34.

249 **birthrates across the world are falling**: See, for example, Stewart Brand, *Whole Earth Discipline: An Ecopragmatist Manifesto* (New York: Viking, 2009), 60–61.

250 *New Republic* **heavy-handedly put it**: H. Fairlie, "Talkin' 'Bout My Generation," *New Republic,* March 28, 1988, 19–22.

250 **"Seniors suck the marrow"**: As cited by James H. Schulz and Robert H. Binstock, *Aging Nation: The Economics and Politics of Growing Older in America* (Baltimore: Johns Hopkins University Press, 2006), 13.

250 **allied to lobbyists for insurance companies**: Ibid., 10–18.

250 **older voters**: Ibid., 205.

251 **voted for Barack Obama**: See 2008 CNN exit poll results at http://edition .cnn.com/ELECTION/2008/results/polls.main.

251 **Kevin Murphy and Robert Topel**: K. M. Murphy and R. H. Topel, "The Value of Health and Longevity," *Journal of Political Economy* 114 (2006): 871–904.

251 **longer life boosts productivity**: D. E. Bloom and D. Canning, "The health and wealth of nations," *Science* 287 (2000): 1207–9.

252 **seminal 1957 paper**: Williams, "Pleiotropy, natural selection, and the evolution of senescence," *Evolution.*

252 **average life span before the dawn**: Robert N. Butler, *The Longevity Revolution: The Benefits and Challenges of Living a Long Life* (New York: PublicAffairs, 2008), 4.

252 **even in the Stone Age**: See R. D. Lee, "Intergenerational Relations and the Elderly," *Between Zeus and the Salmon: The Biodemography of Longevity* (Washington, DC: National Academy Press, 1997).

252 **"grandmother hypothesis"**: See A. R. Rogers, "Economics and the evolution of life histories," *PNAS* 100 (2003): 9114–15, and K. Hawkes, "The Grandmother Effect," *Nature* 428 (2004): 128–29.

253 **Female pilot whales**: See J. R. Carey and C. Gruenfelder, "Population Biology of the Elderly," and S. A. Austad, "Postreproductive Survival," both in *Between Zeus and the Salmon.*

253 **"Nurse bees"**: G. V. Amdam and R. E. Page, "Intergenerational transfers may have decoupled physiological and chronological age in a eusocial insect," *Ageing Research Reviews* 4 (2005): 398–408.

253 **intergenerational transfers**: R. D. Lee, "Rethinking the evolutionary theory

of aging: Transfers, not births, shape senescence in social species," *PNAS* 100 (2003): 9637–42.

254 **"aged dependency ratio":** See Schulz and Binstock, *Aging Nation,* 32–35.

254 **"support burden" in the United States will actually be smaller:** Ibid.

254 **solvent until 2037:** See Social Security Board of Trustees annual report, at www.socialsecurity.gov/OACT/TR/2009/.

254 **modest changes:** Butler, *The Longevity Revolution,* 270–77.

255 **average retirement age:** See M. Gendell and J. S. Siegel, "Trends in retirement age by sex, 1950–2005," *Monthly Labor Review* 115 (1992): 22–30.

255 **labor force participation rates:** Clare Ansberry, "Elderly Emerge as a New Class of Workers—and the Jobless," *Wall Street Journal,* February 23, 2009.

255 **employers regard the penalties:** Schulz and Binstock, *Aging Nation,* 159.

255 **refute the standard bad rap:** See Butler, *The Longevity Revolution,* 245–47.

255 **mentally and physically active:** See Schulz and Binstock, *Aging Nation,* 168.

256 **more than half cited poor health:** Ibid., 154.

256 **too frail to do ordinary work:** As cited in Butler, *The Longevity Revolution,* 474.

256 **By 2017 it will account for:** Will Dunham, "Health care spending surge seen in next decade," Reuters, February 26, 2008.

256 **Robert Fogel has written:** As cited in Butler, *The Longevity Revolution,* 29.

256 **Medicare spending between 1994 and 2004:** K. G. Manton et al., "Long-term economic growth stimulus of human capital preservation in the elderly," *PNAS* 106 (2009): 21080–85.

257 **widening use of costly new medical:** See Schulz and Binstock, *Aging Nation,* 192–98.

257 **In Japan:** Ibid., 191.

257 **improved management of cardiovascular:** See K. G. Manton et al., "Labor force participation and human capital increases in an aging population and implications for U.S. research investment," *PNAS* 104 (2007): 10802–7.

257 **spent more than $100 billion:** Gina Kolata, "Advances Elusive in the Drive to Cure Cancer," *New York Times,* April 24, 2009.

257 **Alzheimer's prevalence:** See "Alzheimer's Disease Facts and Figures, 2007," Alzheimer's Association, at http://www.alz.org.

258 **the number of obese Americans:** "Obese Americans now outweigh the merely overweight," Reuters, January 9, 2009.

258 **Health-care costs for obese adults:** Roni Caryn Rabin, "Obese Americans Spend Far More on Health Care," *New York Times,* July 28, 2009.

258 **greatly elevated risk of dementia:** Shirley S. Wang, "Study Ties Belly Fat to Dementia," *Wall Street Journal,* March 27, 2008.

258 **an astounding 86 percent of U.S. adults:** Y. Wang et al., "Will All Americans Become Overweight or Obese? Estimating the Progression and Cost of the U.S. Obesity Epidemic," *Obesity* 16 (2008): 2323–30.

258 **lowball overweight's health fallout:** See "Body-mass index and cause-specific mortality in 900,000 adults: collaborative analyses of 57 prospective studies," *The Lancet* 373 (2009): 1083–96; and K. F. Adams et al., "Overweight,

Obesity, and Mortality in a Large Prospective Cohort of Persons 50 to 71 Years Old," *New England Journal of Medicine* 355 (2006): 763–78.

259 **In China:** Shirley S. Wang, "Obesity in China Becoming More Common," *Wall Street Journal,* July 8, 2008.

259 **U.S. children are overweight:** J. A. Skelton et al., "Prevalence and Trends of Severe Obesity among US Children and Adolescents," *Academic Pediatrics,* June 26, 2009, epub ahead of print; J. L. Baker et al., "Childhood Obesity—The Shape of Things to Come," *New England Journal of Medicine* 357 (2007): 2325; E. M. Urbina et al., "Youth with obesity and obesity-related type 2 diabetes mellitus demonstrate abnormalities in carotid structure and function," *Circulation* 119 (2009): 2913–19.

259 **overweight children viewed as normal:** Ron Winslow, "Overweight and Overlooked: A Hidden Heart Risk for Kids," *Wall Street Journal,* April 24, 2007.

259 **dismaying prediction:** S. J. Olshansky et al., "A Potential Decline in Life Expectancy in the United States in the 21st Century," *New England Journal of Medicine* 352 (2005): 1138–45.

262 **"longevity dividend":** S. J. Olshansky et al., "In Pursuit of the Longevity Dividend," *The Scientist* 20 (2006): 28–36.

262 **landmark 2005 study:** D. P. Goldman et al., "Consequences of Health Trends and Medical Innovation for the Future Elderly," at http://content.healthaffairs .org.

263 **Richard Miller notes:** R. A. Miller, "Genetic Approaches to the Study of Aging," *Journal of the American Geriatrics Society* 53 (2005): 5284–86.

INDEX